W9-CJZ-655

DECONSTRUCTING
PRODUCT DESIGN

PRODUCT DESIGN

EXPLORING THE FORM, FUNCTION, USABILITY, SUSTAINABILITY,
AND COMMERCIAL SUCCESS OF 100 AMAZING PRODUCTS

WILLIAM LIDWELL AND **GERRY MANACSA**

ROCKPORT PUBLISHERS

BEVERLY MASSACHUSETTS

First published in the United States of America by
Rockport Publishers, a member of
Quayside Publishing Group
100 Cummings Center
Suite 406-L
Beverly, Massachusetts 01915-6101
Telephone: (978) 282-9590
Fax: (978) 283-2742
www.rockpub.com

Library of Congress Cataloging-in-Publication Data
Lidwell, William.
Deconstructing product design : exploring the form, function, usability, sustainability, and commercial success of 100 amazing products / William Lidwell and Gerry Manacsa
p. cm.
Includes index.
ISBN-13: 978-1-59253-345-9
ISBN-10: 1-59253-345-0
1. Industrial design. 2. New products. I. Manacsa, Gerry. II. Title.
TS171.L58 2009
745.2–dc22

2009016039

ISBN-13: 978-1-59253-345-9
ISBN-10: 1-59253-345-0

10 9 8 7 6 5 4 3 2 1

Design: William Lidwell and Gerry Manacsa
Cover Design: Gerry Manacsa and Jill Butler

Printed in China

For our moms . . .

May Lidwell

Dolores Manacsa

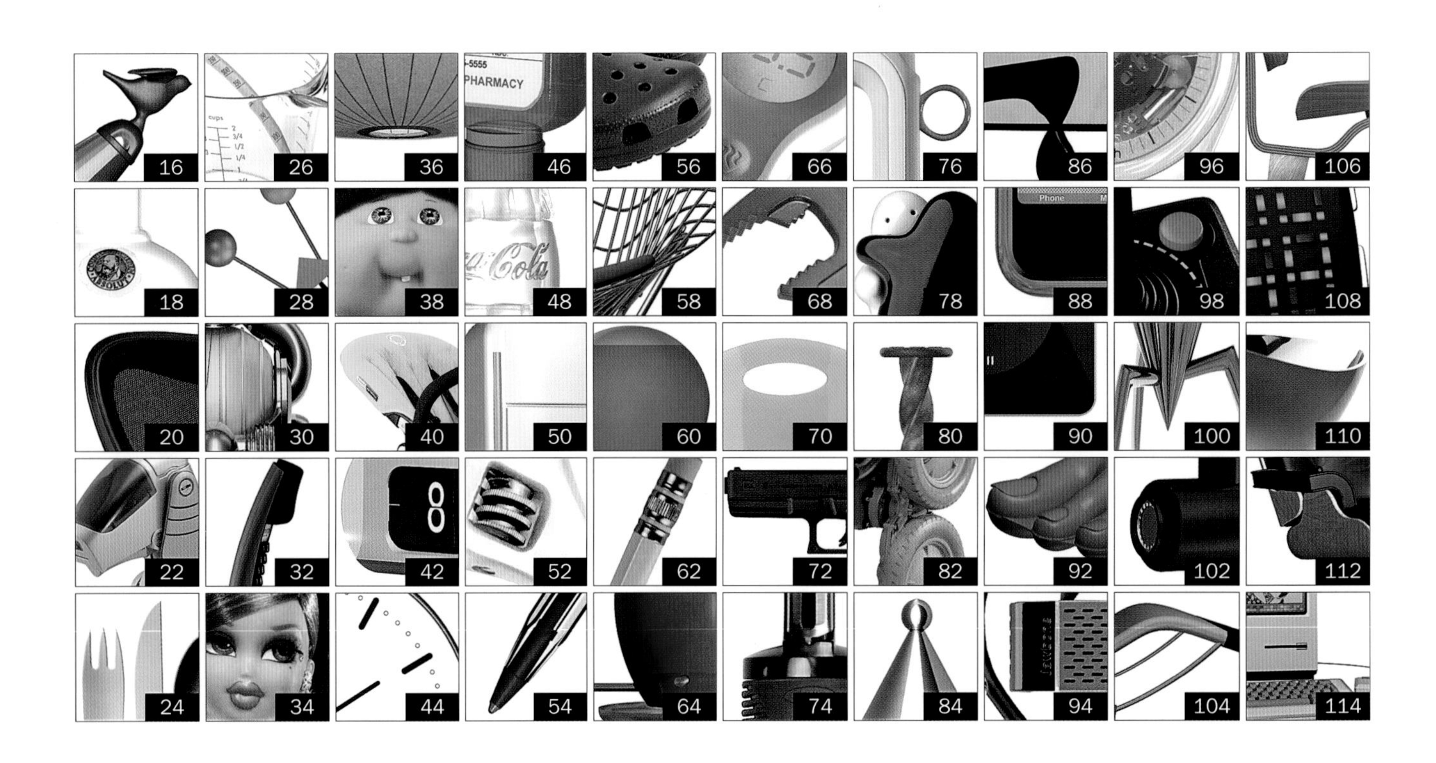

16
26
36
-5555
PHARMACY
46
56
66
76
86
96
106
18
28
38
48
58
68
78
Phone
88
98
108
20
30
40
50
60
70
80
90
100
110
22
32
42
52
62
72
82
92
102
112
24
34
44
54
64
74
84
94
104
114

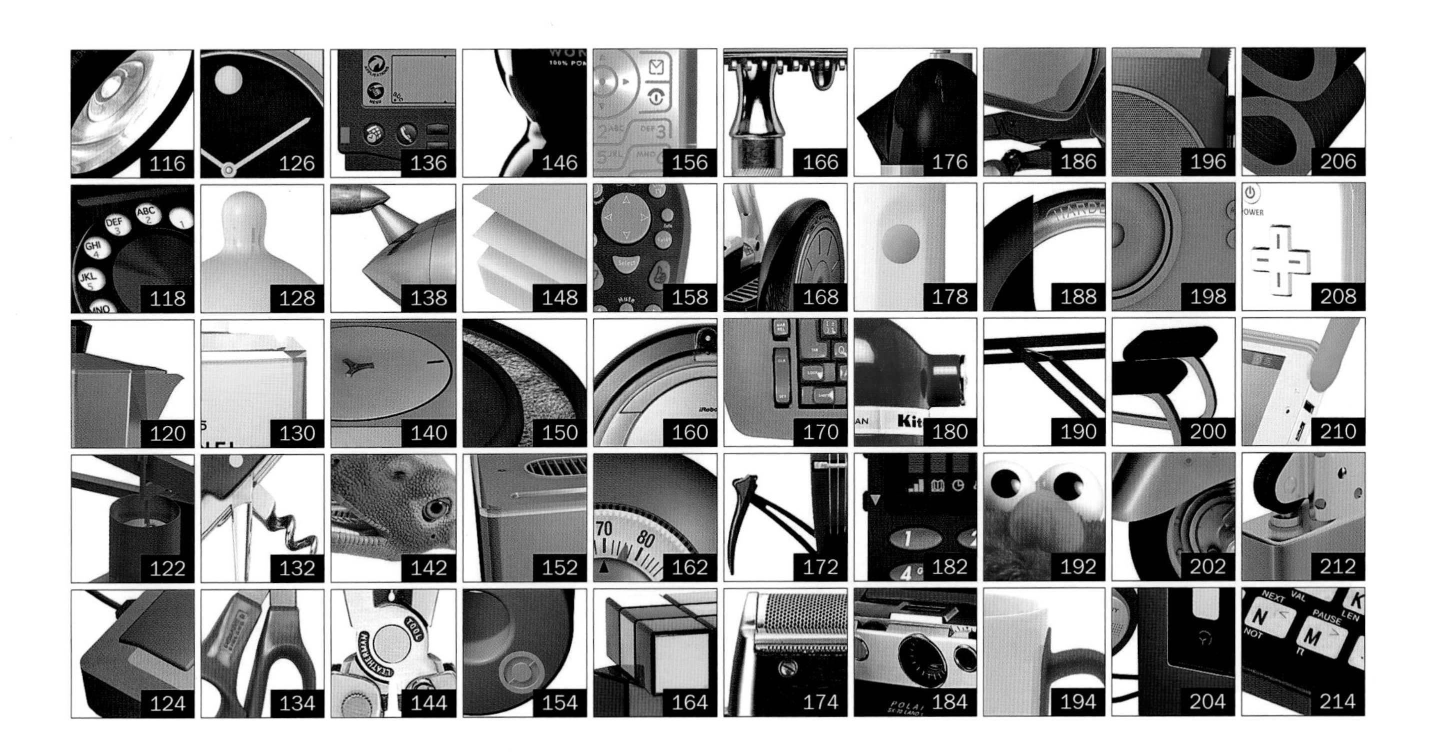

116
126
136
146
156
166
176
186
196
206
118
128
138
148
158
168
178
188
198
208
120
130
140
150
160
170
180
190
200
210
122
132
142
152
162
172
182
192
202
212
124
134
144
154
164
174
184
194
204
214

Stop this day and night with me, and you shall possess the origin of all poems,
You shall possess the good of the earth and sun, (there are millions of suns left,)
You shall no longer take things at second or third hand, nor look through the eyes of the dead, nor feed on the specters in books,
You shall not look through my eyes either, nor take things from me,
You shall listen to all sides and filter them from your self.

WALT WHITMAN

THE PURPOSE OF THIS BOOK is to get you to think hard about design—to *deep process* it—and in so doing, to make you a better design thinker.

To this end a number of strategies are employed: emergent themes across 100 products hint at heuristic definitions of "good design" and "bad design"; tacit references to principles, terms of art, and historical movements evoke questions, and in some cases will require additional research; semantic icons stimulate analogical thinking and ideation; qualitative ratings across five dimensions—aesthetic, function, usability, sustainability, and commercial success—provide an objective framework for apples-to-apples comparison of products, and implicitly challenge you to evaluate both our ratings and the dimensions used to asses the merits of a design; multidisciplinary commentary offers varied and often contradictory perspectives to broaden thinking and highlight elements relevant to different design disciplines; when available, a quotation from the designer or other individual close to the product offers a sense of the product's design inspiration and philosophy; and critical analysis explores the salient aspects of the product design, which ranges from form and function to marketing and socio-historical context.

If you find the prospect of grappling with all of this is exciting, you will find much to mine in this book. And your efforts will be rewarded with insights and techniques that can have an immediate impact on the quality of your design strategy and practical work.

If the prospect of this level of engagement is less than exciting to you, do not despair, as the book functions on other levels: a primer on design literacy and history, a series of case studies, a book of criticism (in the literary sense of term), and a philosophic exploration of the meaning of design. The book can be read at these different levels on different occasions, in linear or nonlinear fashion, to suit your particular tastes at the time.

Finally, a word about the products in the book. One hundred products were selected based on two criteria: 1. Does the product exemplify "good design" in at least one significant and edifying respect? and 2. Is the product fruitful in its capacity to illustrate one or more key principles of design? Purists will no doubt cringe that such classics of product design as the LCW Chair and Phonosuper SK 4 share the volume with the likes of the Bratz doll and MoneyMaker pump, but that is, in part, the point—to make them cringe, and then to unapologetically inquire as to why one product is deemed classic and another banal. You may find that many heralded classics of design bear a striking resemblance to the finery of the emperor's new clothes, whereas other lesser-known and more mundane products are true classics awaiting their proper acknowledgement. We invite you to consider both possibilities, to internalize the various ideas and opinions presented on each page spread, and then, as appropriate, to "filter them from your self."

William Lidwell
Gerry Manacsa

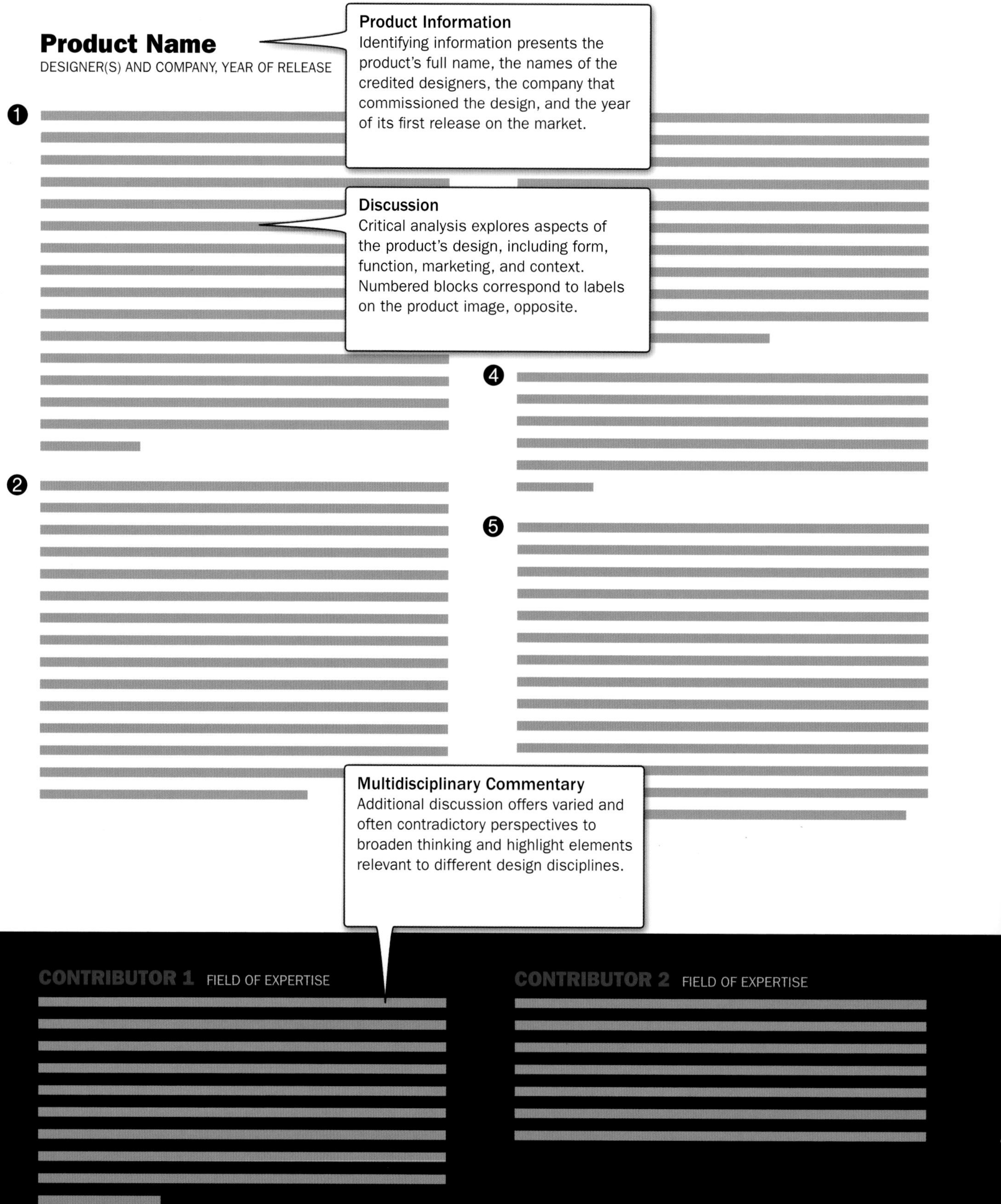

Product Name

DESIGNER(S) AND COMPANY, YEAR OF RELEASE

Product Information
Identifying information presents the product's full name, the names of the credited designers, the company that commissioned the design, and the year of its first release on the market.

Discussion
Critical analysis explores aspects of the product's design, including form, function, marketing, and context. Numbered blocks correspond to labels on the product image, opposite.

❶

❷

❹

❺

Multidisciplinary Commentary
Additional discussion offers varied and often contradictory perspectives to broaden thinking and highlight elements relevant to different design disciplines.

CONTRIBUTOR 1 FIELD OF EXPERTISE

CONTRIBUTOR 2 FIELD OF EXPERTISE

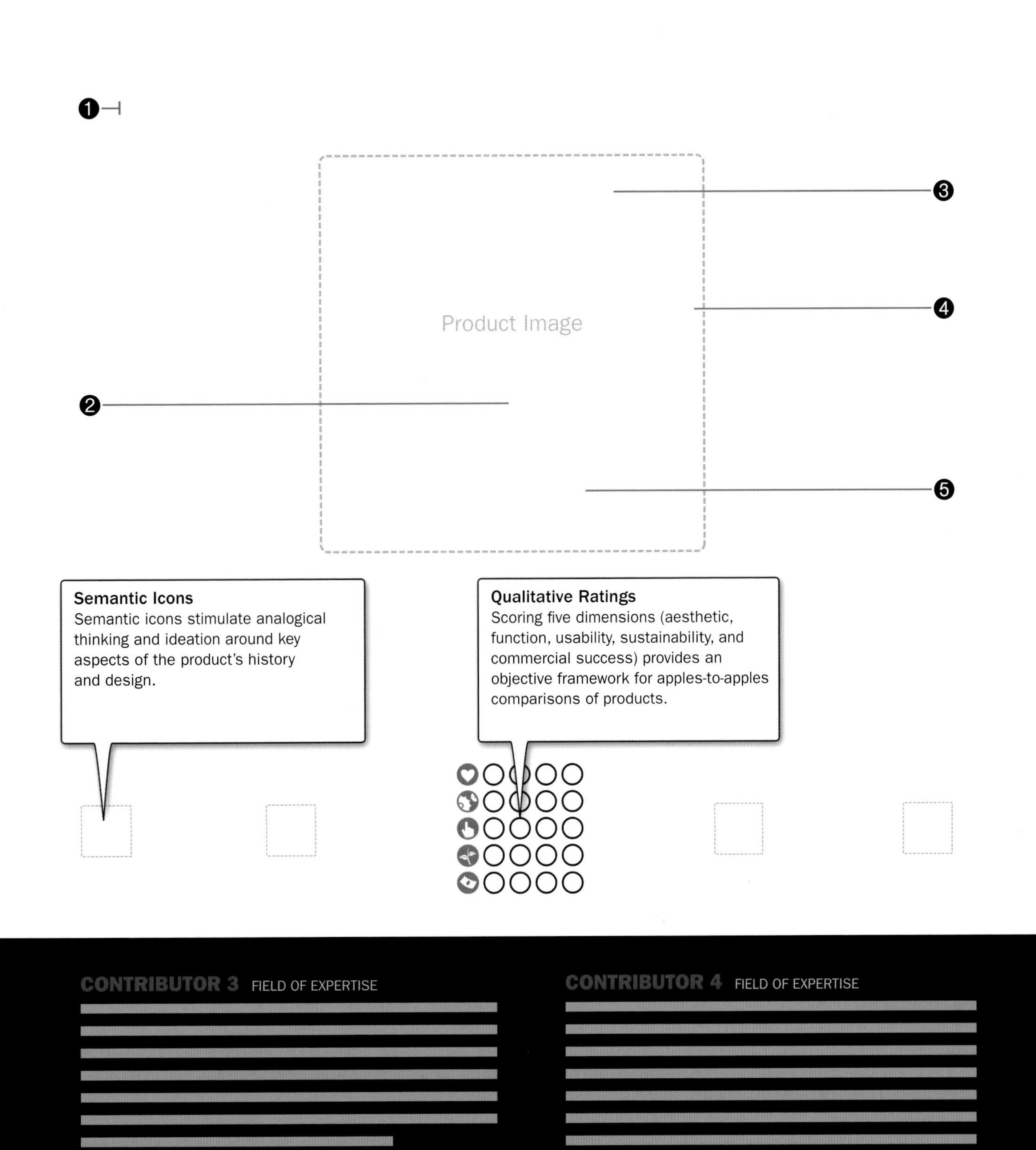
Product Image
Semantic Icons
Semantic icons stimulate analogical thinking and ideation around key aspects of the product's history and design.
Qualitative Ratings
Scoring five dimensions (aesthetic, function, usability, sustainability, and commercial success) provides an objective framework for apples-to-apples comparisons of products.
CONTRIBUTOR 3 FIELD OF EXPERTISE
CONTRIBUTOR 4 FIELD OF EXPERTISE

PRODUCTS

9093 Kettle

Michael Graves for Alessi, 1985

❶ When Alberto Alessi approached Michael Graves to produce a “functional” kettle, the second of its designer kettle series, he did so with a well-defined set of requirements, including the need for it to whistle. Graves had to work within these rigid requirements to produce a final design — one that ultimately became Alessi’s best-selling product and spawned a family of related items. What is it about this playful kettle, priced high for its category at over $100, that has so intrigued consumers? Michael Graves comments: “Architectural and product designs have a narrative capacity — you can start to tell a story about them and imagine a lot of things. The Alessi bird kettle has a personality, with its simple geometry. Its dots on the bottom are red to signify heat as it’s placed on the stove. And the shape of the grooved handle, which is blue where it was cool to touch, and, of course, a bird whistles.”

❷ The conical stainless steel kettle possesses an unremarkable geometry. It is the ornamentation — bird, handle, rivets — that differentiates the kettle and endows it with personality. A basic color code is applied: red elements get hot, blue elements remain cool. Its form is at once playful, approachable, and sophisticated, a signature style of its postmodern designer, and a combination largely responsible for its mass-market appeal. One wonders, however, if its success has more to do with its marketing, with its celebrity architect association, than its blue handle or whistling red bird, enabling its premium price point and transforming an otherwise commodity kettle into a kind of Veblen good, where the high price drives the demand and the physical design is secondary.

❸ The handle remains cool at all times, true to its color code. The grip is positioned along a circular metal rod, just off center toward the back, extending up and over the lid. This position facilitates pouring but somewhat obstructs removal of the lid. The color and slight finger-formed depressions nicely afford gripping. The front and back of the grip are terminated with red balls, indicating that the exposed metal just beyond them heats up with the body.

❹ The most eccentric feature of the kettle is the polyamide red bird perched at the mouth of the spout, a cap that creates a disappointingly nonbirdlike whistle when water boils. Still, the color and iconic rendering of the bird define the product. The bird must be removed to pour water. Aside from the obvious burn hazard, removing the cap every time is enough of a hassle that it is often just left off and eventually lost. A hinged cap, a feature offered on its less expensive Target progeny, would be a welcome modification.

❺ The exposed rivets near the bottom of the base are an interesting aesthetic choice. Their presence is clearly ornamental — reminiscent of the rivets once used to secure base to body in older kettles, though here they just secure the kettle as an Art Deco object. The rivets do offer a subtle aesthetic counterweight to the top-heavy ornamentation, and they convey a sense of strength and stability, echoing the solidity of the base and the kettle’s high-quality stainless-steel construction.

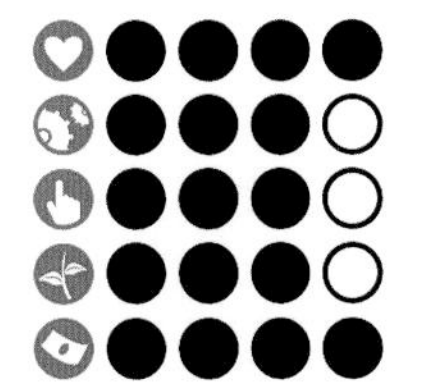

Absolut Vodka Bottle

Carlsson and Broman, 1977

❶ Derided as resembling a plasma bottle by early critics, the Absolut Vodka bottle achieved commercial success and design recognition soon after its introduction to America in 1979. Its subdued design is an actualization of Gunnar Broman's "when all others are screaming, you must whisper" philosophy—a reference to the loud, if not garish, bottles of competing brands of the time. The Absolut bottle has long since achieved iconic status, no doubt aided by its role as centerpiece in one of the longest-running advertising campaigns of all time: 25 years and 1,500 advertisements all built around the bottle.

❷ Its form was inspired by an antique Swedish medicine bottle, a nod to vodka's historical use as a medicinal and a hint as to the purity of product, which was considered its primary differentiator. These themes are further reinforced through the use of clear, uncolored glass unobscured by traditional labeling. This see-through aspect of the design proved a significant point of controversy. Distributors raised concerns that the bottle would either act as a magnifying glass, enlarging competitor labels, or disappear on the shelf altogether—concerns that were ignored and ultimately proved to be unfounded. The neck of the bottle was lengthened from the original to add height and improve its gripping and pouring affordances. A silver plastic cap gently blends with the bottle and eliminates issues of breakage and sediment associated with cork. A small glass rim at the foot of the neck reinforces the joint and adds a questionable ornament that subtly increases visibility. The domed bottom, or *punt*, increases base strength and visual interest in profile, and gives the bottle a classic hand-blown appearance.

❸ The seal—intended as a joke until it received positive reviews, at which time it became a feature—bears the likeness of Lars Olsson, a Swedish innovator of distillation methods in the late nineteenth century. It is pure ornament, but it adds a historical patina that, like the script, creates interesting interplay with the otherwise modern aesthetic.

❹ The name is boldly presented in a flavor of Futura, setting a clean modern tone. The type is colored blue to reinforce purity. The brand Absolut is cleverly exploited as a double entendre in both the product name and advertising campaigns (e.g., Absolut New York), making it at once descriptive and declarative. Given that Sweden was not known internationally for their vodka at the time, and given that the target market was America, the emphasis and positioning of "Country of Sweden" was likely to emphasize that the source of the vodka was somewhere other than a country in the Soviet Union. The script was written by hand by Kotte Jönsson, and offers a classic, sophisticated countermelody to the Futura.

JILL BUTLER GRAPHIC DESIGN

Not knowing a lot about wine, I often make purchases based solely on the visual design of a wine bottle. Based on that criterion, I would not be drawn to purchase a bottle of Absolut Vodka. The shape of the bottle and the logo result in a simple, clean appearance. But the two sentences of centered text in the script typeface are so prominent that, overall, the bottle looks a lot less modern than its shape and logo. Because of this text, the bottle reminds me of a

BROCK DANNER ARCHITECTURE

Exemplifies the power of simplicity and focus in design. Whether we discuss it as a modernist dematerialization of ornament, or as an unconscious suggestion of medicinal healing contents, the design has successfully become a symbol of spirits nevertheless. It achieved its success using a simple formula—simplicity of form, transparency, and a bold universal typography balanced against a field of script.

1
2
3
4
ABSOLUT®
Country of Sweden
VODKA
This superb vodka
was distilled from grain grown
in the rich fields of southern Sweden.
It has been produced at the famous
old distilleries near Åhus
in accordance with more than
400 years of Swedish tradition.
Vodka has been sold under the name
Absolut since 1879.
40% ALC./VOL. (80 PROOF) 1 LITER
IMPORTED
PRODUCED AND BOTTLED IN AHUS, SWEDEN
BY THE ABSOLUT COMPANY
A DIVISION OF V&S VIN&SPRIT AB.

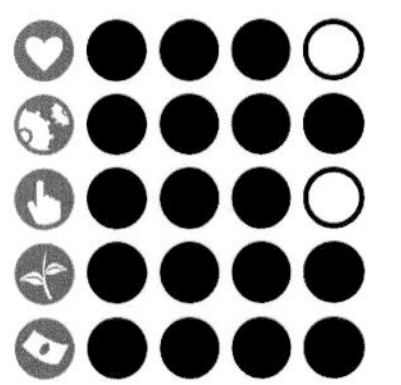

Aeron Chair

Bill Stumpf and Don Chadwick for Herman Miller, 1992

❶ With radical emphasis on function over form, the Aeron revolutionized both office ergonomics and office sociology. Gone are the days of one-size-fits-all chairs and office castes where title and position are indicated by the size and padding of one's seat. It was the Aeron that made it so. The product of substantial research conducted by Herman Miller on how people *really* sit versus how they *should* sit, the Aeron is, in a sense, born of pure inductive design, enabling the abandonment of superstitions and vestigial conventions about how chairs ought to be. The result is an object that disappears when used—an object that, rather than seeking continuous notice and recognition, seeks only to fade from consciousness and become one with its user. Bill Stumpf comments: "True comfort is the absence of awareness. If your shoes are comfortable, you're not aware they're on. If the water is pure, you can't taste it. Similarly, when a chair is a perfect fit for your body, it becomes 'invisible,' and you're not aware of it at all."

❷ The chair comes in three sizes (small, medium, large) to better accommodate the full distribution of human dimensions. The remaining variability is addressed through a series of controls, all of which are readily accessible from the seated position, except for the lumbar support. The frames are curvy and biomorphic, spanned with an elastic mesh called pellicle, which is impressively breathable, durable, and easily cleaned. The chair is black with an occasional hint of metal, though modern versions have marginally expanded the palette to a small set of neutrals. The use of a limited neutral palette reinforces the relatively high price point, but, as with Henry Ford's Model T, also limits consumer appeal and invites competitors.

❸ The armrests are independently adjustable vertically and horizontally, enabling users to easily lock themselves in the chair. As with the seat, the armrests have a waterfall front edge. Interestingly, an implicit claim of the design—that is, spanned pellicle is superior to traditional padding for support and comfort—is contradicted by using traditional padding in the armrests. It is not clear why the rationale of using spanned pellicle on the back and seat does not apply equally to the armrests, and the cost is incoherence in the design.

❹ The seat is effective at accommodating variable seating behaviors (e.g., lip sitting versus leaning back). The waterfall front edge of the seat reduces problems of circulation under the thighs, though it would benefit from a greater angle of attack. The span is limited in its capacity to offer varying degrees of support across the uneven surface of the human buttocks, making it less comfortable to many than its more traditional padded counterparts. That said, the aeration afforded by the pellicle is generally a good trade, as most will happily endure numb buttocks to avoid sweaty thighs.

❺ The mechanicals are largely exposed, giving the supporting structure an exoskeletal appearance. Take an organically designed object, paint it black, give it five legs, expose its mechanicals, subtract any semblance of padding, and you get a chair that is decidedly uninviting—it looks hard, arachnoid. This lack of *prima facie* appeal is quickly overcome, however, when a person sits in an Aeron, a requirement that is undoubtedly the greatest challenge of marketing this product.

ALAN COOPER TECHNOLOGY DESIGN

The Aeron changed lots of minds about how industrial design can drive sales by transforming something as quotidian as an office chair into a coveted perquisite. Its swoopy shape and tactile materials begged to be touched, but what really

BROCK DANNER ARCHITECTURE

This chair is a product of evolution in ideology and societal sensitivity. The task chair is the seat of the worker, once an enslaved hand, then a numerical calculation of time and efficiency, and only recently as a human being seated

1
2
3
4
5

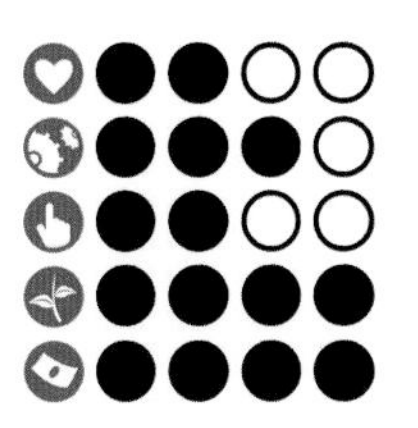

AIBO ERS-110

Hajime Sorayama and Sony Corporation, 1999

❶ AIBO, derived from the Japanese word *aibou* meaning "partner," and secondarily serving as an acronym for *artificially intelligent robot*, began as a skunkworks project at Sony around 1994. The question from its conception primarily regarded its function—What function would this robot dog serve? Takeshi Yazawa, vice president and general manager of Sony Entertainment Robot America, comments: "We had lots of arguments about whether AIBO should do something or not. But in the end, we all agreed that it wouldn't do anything useful at all. It would be a pet. AIBO loves you, you love AIBO, and that's it. In many ways it would have been easier for us to focus on trying to make a robotic washing machine or robotic newspaper delivery machine, or something functional like that, but in the twentieth century, we [Sony] had developed a lot of devices associated with specific functions. We said to ourselves, 'It's the twenty-first century now; this is our opportunity to move away from that.' Of course, we got lots of questions within Sony about this decision. Mr. Idei [Sony Chairman and CEO] asked the same question that everyone asked when they saw the first prototype—'But what can AIBO *do*?' But then his second question was, 'Can we sell this?'"

❷ Unlike other entertainment robots like Furby and Pleo, the AIBO unapologetically embraced its robotness, a tribute to the robot fetish of artist and designer Hajime Sorayama. Attempts were made to tweak and refine the aesthetic of subsequent models, but they were banal and directionless. It is clear that when the product lost its impassioned robotic visionary, it lost its way. Sorayama's original design continues to be one of the most endearing robotic forms ever developed.

❸ AIBO looks and moves like a robotic puppy, though with the clunkiness one would expect from the first entertainment robot. Movements and reactions are painfully slow. Physical behaviors are canimorphic—for example, playing with a ball, plausible physical responses such as head tilting and tail wagging, exploring, learning, and sleeping—and emotional states are more anthropomorphic—for example, joy, sadness, anger, surprise, fear, and discontent are indicated through body language, auditory tone combinations, and eye lamp patterns. The result is that many users develop a deep affection for their AIBOs, often naming them, speaking to them as they would a pet, and in some cases even attending AIBO gatherings where the robotic pooches entertain their proud owners with coordinated line dancing. Despite this, the AIBO remained distinctly robotic. It speaks "robot language" akin to the tones and beeps of R2D2. No attempt is made to silence its servos or hide its cooling vents; rather, these elements are integral to the design. If users should forget at times that they are dealing with a robot fashioned to look like a dog, the fact that the memory stick and batteries are inserted into the anal region of the bot helps snap everyone back to reality. The AIBO ERS-110 was an impressive first-generation product, but it had an equally impressive price point that limited its audience to über-affluent über-geeks. Never satisfactorily answering the question, "What function would this robot dog serve?" exacted a high price in the form of scope creep. New capabilities were added with each model, but to what end? What were the requirements driving the design? Without a well-defined purpose, the AIBO evolved into an unaffordable, tech-bloated R&D indulgence that enjoyed little commercial success beyond its initial release.

❶
❷
❸

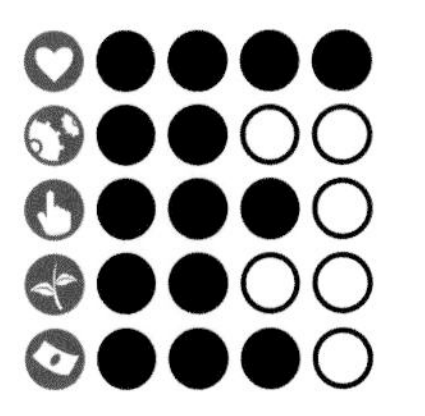

AJ Cutlery

Arne Jacobsen, 1957

1. The cutlery is a product of Jacobsen's theory of integrated design and architecture, created as part of his SAS Royal Hotel commission, one of many objects ranging from door handles to ashtrays created specifically for the hotel. The modernist implements, however, were poorly received by guests and replaced by traditional cutlery a short time after their introduction. It was not until their selection by Stanley Kubrick to appear in the movie *2001: A Space Odyssey* that the space-age cutlery enjoyed a revival, eventually emerging as an icon of modernist design. Just before his death in 1971, Jacobsen summarized his general approach to design: "First of all, it is the proportions, and next comes the texture—that you do not mix the wrong materials together—and out of this, of course, comes the colors, and altogether the general impression."

2. Each piece of the set is minimal and functional, forged from single pieces of stainless steel, seamless, and sans ornament of any kind, which makes for easy washing. Their feel is heavy and precise, like ruggedized medical instruments. Indestructible. Curves and bends are so gentle that the cutlery appears naturally formed, as if molten metal was halted mid-drip and then polished to matte finish. The aesthetic is at once modern and organic. For those with modern tastes, the cutlery represents perfection—simple, symmetrical, and functional. They are deemed far less than perfect by those with a penchant for shoveling food into their mouths—a decidedly unhealthy American style of consumption—highlighting the principal user complaint: the scooping areas of the fork and spoon are small relative to their more traditional counterparts. Given the worldwide rising trend in obesity, however, one wonders how user-centered the design of cutlery should be.

3. The spoon is the most sculptural implement of the set. The scoop is a gentle recess and expansion of one end of the handle, set apart by a gentle bend upward. The spoon has the appearance of an elongated ladle. The shape makes inserting the entire scoop into the mouth impracticable, and is a point of contention among those accustomed to doing so. As with the knife and fork, the spoon is uncharacteristically strong, capable of scooping firm ice cream without fear of bending.

4. The knife looks like a scalpel but is blunt by comparison—it is really more wedge than knife, but cuts well nonetheless. A gentle dip from the top line provides users a place to put their index finger to apply additional pressure when cutting, and provides a visual indication of where the blade begins. Given the relative bluntness of its edge, it is interesting that the knife departs from the uninterrupted lines of the knife and fork. Perhaps that would have been too austere, even for Jacobsen.

5. Make no mistake, this fork is for sticking into food and inserting it into the mouth only—there is minimal surface for scooping and the short, teethlike tines make twirling noodles futile. If one seeks a stylish means to slow the rate of eating, this is an ideal choice. The two outer tines extend naturally from the lines of the handle and contain a third, creating a modern trident. The tines are rigid and sturdy and do not bend like other forks, when separating a tough steak or abused by a dishwasher—so much so, in fact, that those inclined to use their fork to mince food will find that the help of a knife is usually not needed.

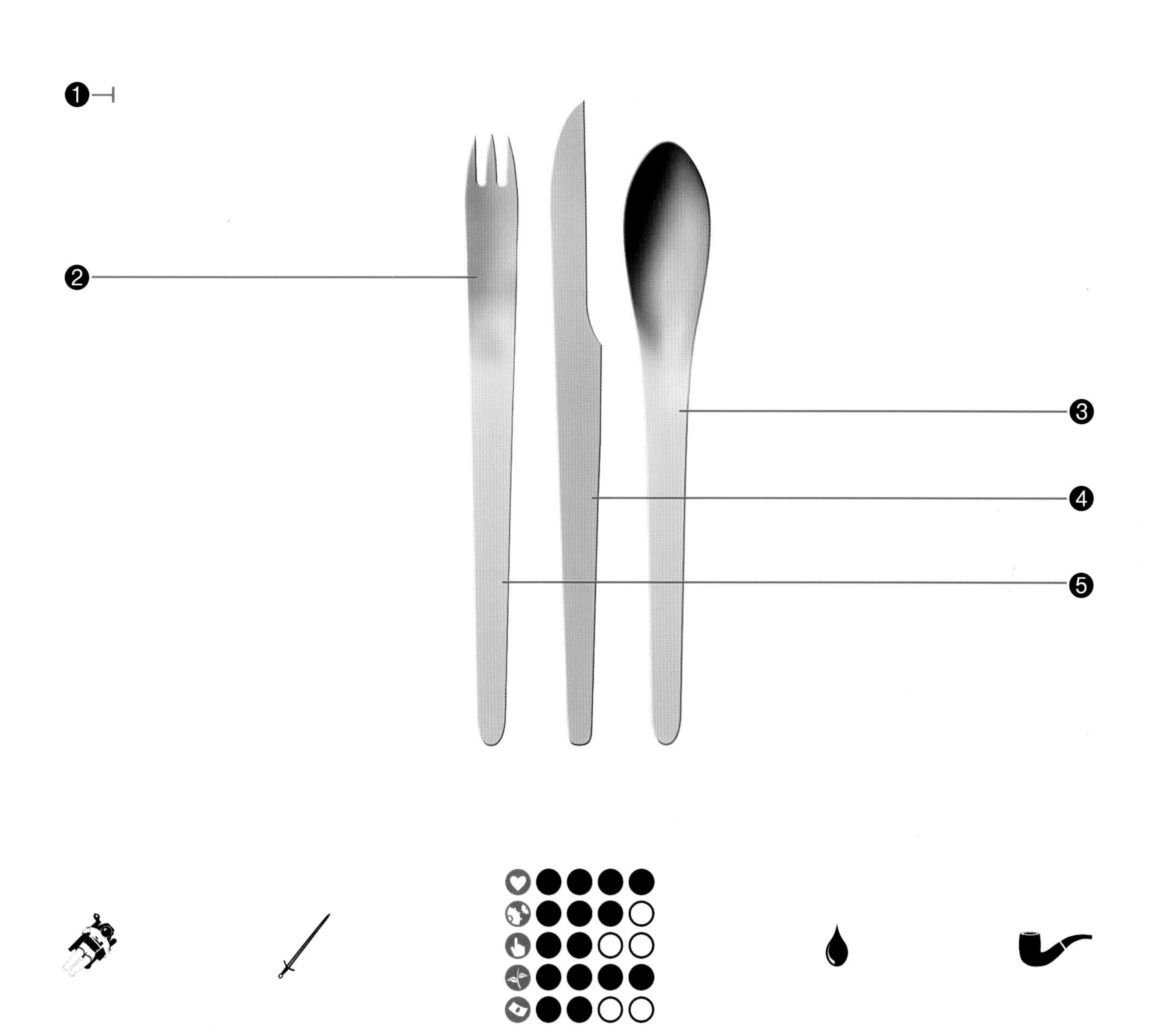

Angled Measuring Cup

Bang Zoom Design and Smart Design for OXO, 2001

❶ When the toy-designing cousins Michael Hoeting and Stephen Hoeting submitted their prototype of an angled measuring cup to OXO, the genius of the design was immediately apparent. The question, however, was whether the cup was a gimmick or actually addressed a relevant user problem. Alex Lee, CEO of OXO International, comments: “We went and got a bunch of users and asked them the question, ‘What is wrong with your measuring cup?’ And they would say, ‘Since it is made of glass, it breaks if it drops; when my hands are greasy they are slippery; when I heat things up they get hot.’ And we’d say, ‘Okay, anything else?’ And they’d say, ‘No, we are happy otherwise.’ So we’d say, ‘Show us how you measure.’ They’d pour, bend down, look at it. Pour some more, bend down, look at it. Four or five times. Nobody mentioned this as a problem, because this is an accepted part of the process of measuring. And we are happy when we see this kind of problem—this clear inefficiency that nobody articulates.”

❷ The form is functional but busy, remarkably uncompelling for an OXO product. The key innovation and attractor is the angled elliptical ramp inset from the inner walls that displays the common units of measure. This enables users to easily gauge the volume from the top, reducing parallax error and eliminating the up-and-down gymnastics normally associated with achieving a precise measurement. The innovation is not without costs—the ramp impedes stirring within the cup and exaggerates error when not held perfectly level—but these are quibbles overshadowed by a significant improvement in usability. The handle is what one would expect from OXO: imminently grippable.

❸ Ramp quantities are presented red on white in cups and ounces. Contrast is good, but typography and layout are not—the markings appear slapped on, varying in size, weight, and alignment, and do not instill confidence in their precision. Perhaps because the cup is made of transparent plastic, the designers felt compelled to add traditional measurement markings on the side. This unfortunate decision undermines the innovation in the design, contradicts its intended use, and adds considerable noise to the form. These markings are relatively low contrast, and the vertical orientation of the lines and numbers distracts from the angular element, which is the product’s key differentiator. The measuring cups come with a sticker on the base that reads, “Angled surface lets you read measurements from above.” A sticker like this is required because the side measurements afford improper use. Eliminate the competing affordance and there is but one intuitive way to use the cup. The design would be well served to follow its stainless-steel counterpart and lose the side markings—and with it the sticker.

❹ The handle is firm and grippy, extending the full height of the cup. It offers ample surface area for holding and doubles as a supporting buttress when the cup is set down. The handle is connected to the cup by an ergonomically concave thumbwell, perfectly sized and textured to support controlled pouring and minimize slippage.

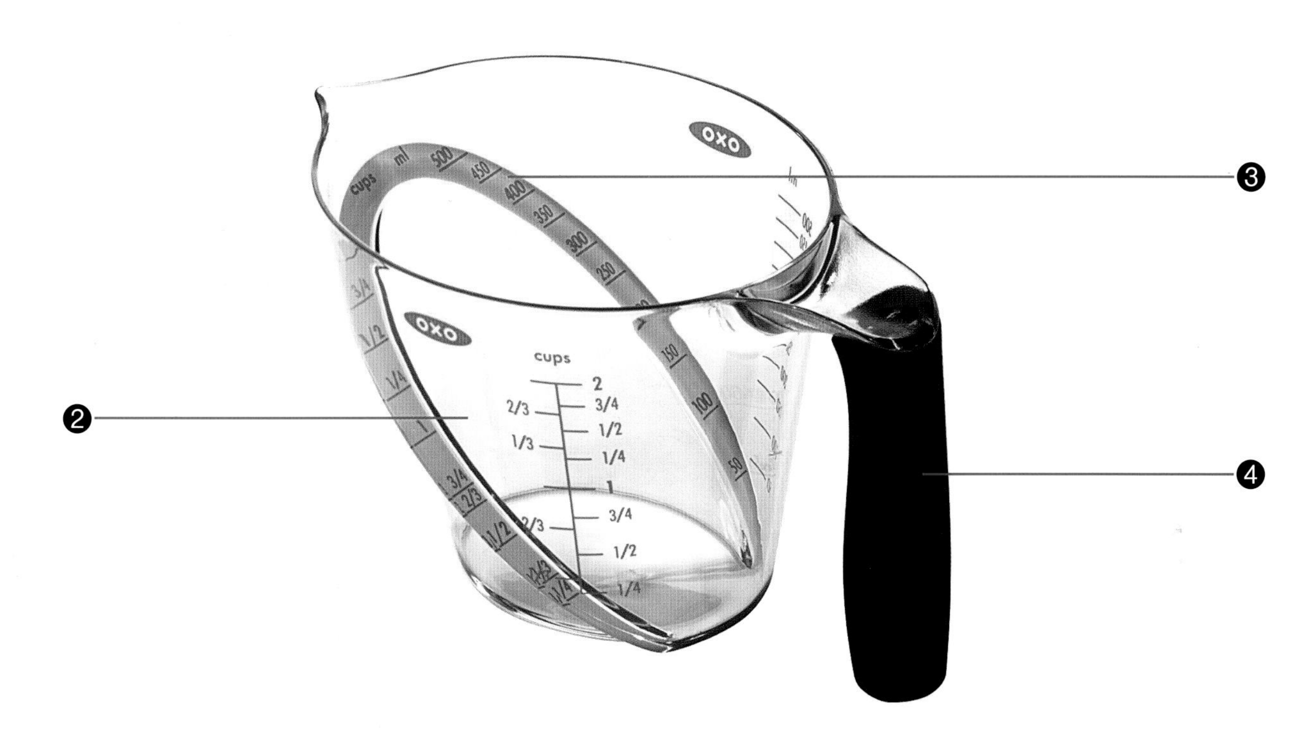

OXO
cups
2
3/4
2/3
1/2
1/3
1/4
1

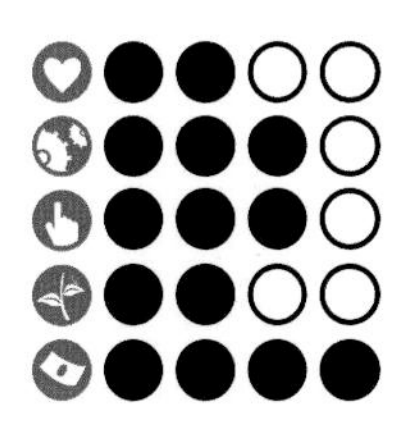

Ball Clock

Irving Harper and George Nelson Associates, 1948

1. Inspiration for great design can come from different places — necessity, opportunity, analogy, and sometimes where you least expect it. George Nelson comments: "Noguchi came by, and Bucky Fuller came by. I'd been seeing a lot of Bucky those days, and here was Irving and here was I, and Noguchi ... he saw we were working on clocks and he started making doodles. Then Bucky sort of brushed Isamu aside. He said, 'This is a good way to do a clock,' and he made some utterly absurd thing. Everybody was taking a crack at this ... pushing each other aside and making scribbles. At some point we left — we were suddenly all tired, and we'd had a little bit too much to drink — and the next morning I came back, and here was this roll [of drafting paper], and Irving and I looked at it, and somewhere in this roll there was a ball clock. I don't know to this day who cooked it up. I know it wasn't me. It might have been Irving, but he didn't think so ... it could have been an additive thing, but, anyway, we never knew." Subsequent discussions with Irving Harper suggest that it was he that designed the clock, and that Nelson's account was a colorful embellishment of history. Either way, perhaps the lesson is that inspiration for great design comes not only from the work of the solitary genius motivated by humble circumstances or inspired by some great musical score, but also from great designers working together to solve interesting problems.

2. The form is a simple configuration, six lines intersecting at midpoint terminated by spheres, making the clock look like a cross between a spoked wheel and molecular model. The oddly shaped hands and mix of colors give the clock a cartoonish quality, a postmodern statement appropriate at the time of its development, but an aesthetic that has grown tired and today seems flippant, like a rushed student project.

3. The lines are constructed of metal rods, providing a rigid and structurally minimal support for the spheres. The lines of the clock in concert with the spheres create an abstract, yet playful aesthetic. Unfortunately, the lines also add visual noise, conflicting with the hands and compromising readability. Accordingly, the clock is at its strongest when the rods are painted the same color as the wall on which it is mounted, minimizing noise and emphasizing the hands and spheres.

4. The hands are the most questionable elements of the clock, and are often the first sacrificed by copycats and derivatives. One could argue that the oversized curvy arrow and oddly inset ellipse somehow help the eyes process location and orientation, but surely the shapes of the hands confuse as much as they aid, and are more likely the result of playful indulgence followed by a lack of editing.

5. The spheres are painted wooden balls, replacing numbers as the hour markers and serving as the primary innovation of the clock. The design is based on the premise that people tell time by looking at the position of clock hands versus the numbers. This turns out to be generally correct for hand positions at hour and quarter-hour positions, but far less true for other intermediate positions, trading readability for form.

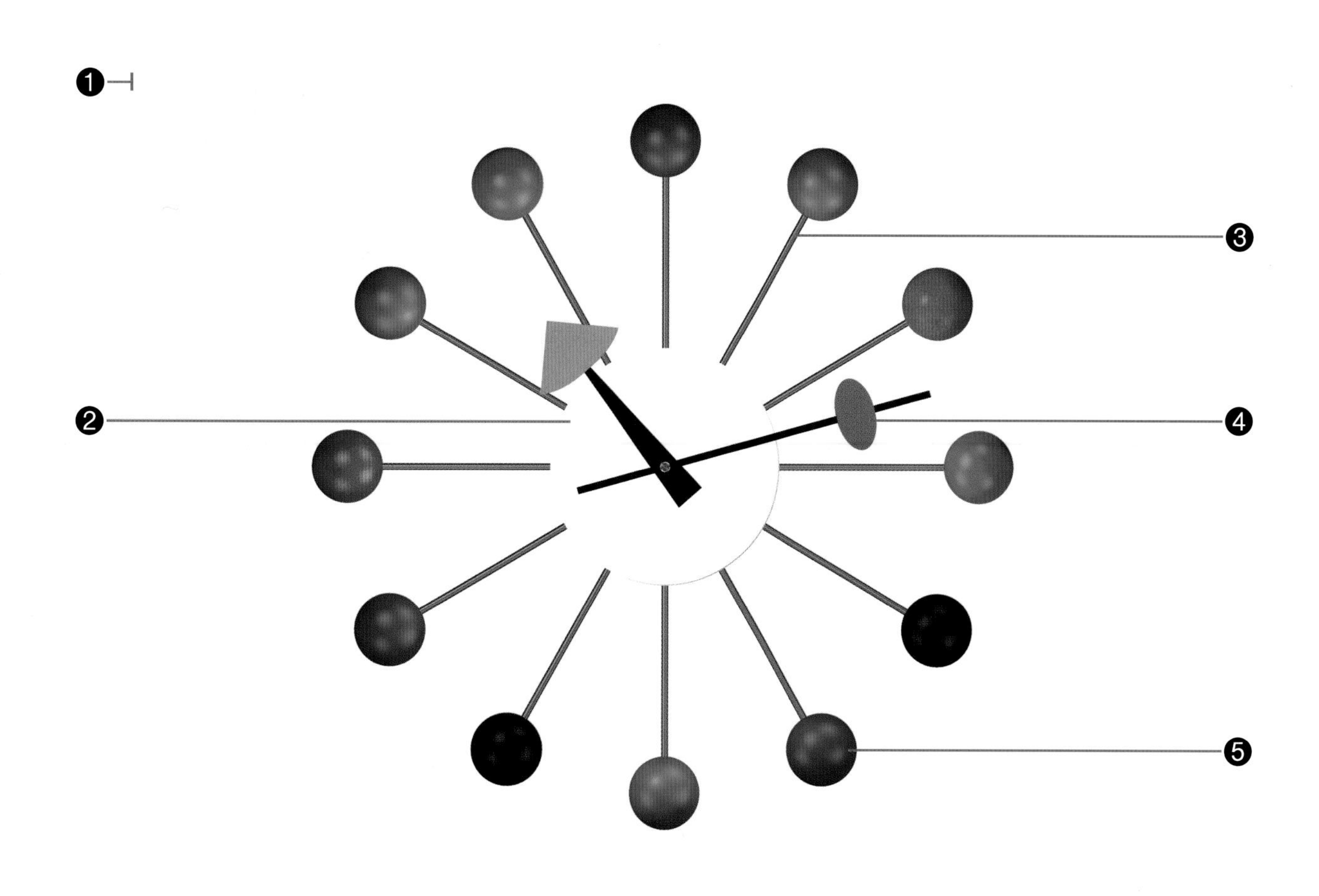
1
2
3
4
5

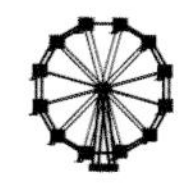

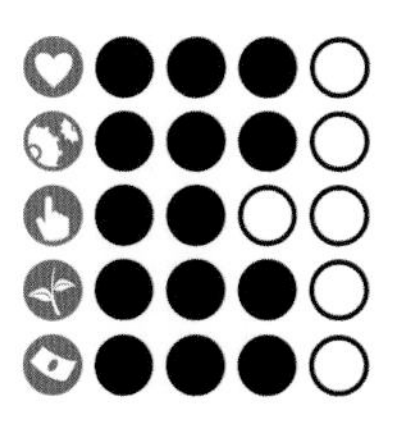

Ball Vacuum Cleaner

James Dyson, 2005

1. The Dyson series of vacuum cleaners has risen from obscurity and humble beginnings to become a dominant force in the premium vacuum cleaner market. The products have a unique aesthetic, appearing as hybrid vacuum cleaner-robots, embodying a design approach perhaps best characterized as *elucidative design*—an approach more about edifying users in the differential value of the product than seeking either a purely functional or purely beautiful form. James Dyson comments: "I very rarely talk about form because I think performance is much more important—that form has a role because it's often people's first impression. But I'd rather say that when you look at the object, you've got to be able to tell what it is and what it does, and be taught something, and be excited in some way. And that isn't necessarily form following function, but rather something probably much more complex than that. So, for example, part of the reason for the clear bin is so you can see the technology inside, and you can see how it works. And not concealing the pipes on our products isn't because we want to do 'form follows function,' but because I think it's important that people understand how they work."

2. The form is bulgy and mechanical, unrefined looking, and is dominated by the cyclonic head and large ball, both highlighted in yellow to capture attention. The bin, with its metal innards exposed and sandwiched between the highlighted elements, is next in the reveal. Sequence of presentation is spot on—differentiated features are presented first. The predominantly plastic body seems at odds with the high price point, making distinctively cheap-plastic creaking noises when flexed, but the trade-off clearly regards weight (and presumably cost) concerns.

3. The vacuum cleaner uses a series of cyclonic separators to remove particulates from air without the use of filters, a technology modeled after large-scale separators used in sawmills. Separation occurs by spinning vacuumed air at high velocity in a spiral pattern, beginning at the wide top of the cyclonic chamber and ending at the narrower bottom. This forces particles in the rotating stream outward, striking the outside wall where they fall to the bottom and exit. Cleaner air spirals upward toward an exhaust at the top. The method is masterfully applied and suction is, indeed, constant after long-term use. If only Dyson could make a silent vacuum cleaner!

4. The cleaner is bagless, storing dirt in a removable transparent bin, a point of significant controversy upon its introduction and a feature panned by focus groups and retailers—prevailing wisdom was to conceal, not highlight, collected dirt. Dyson, however, was firm in his belief that people would want to see the cyclones in action. The bin remained clear, and the decision was ultimately vindicated by sales. The bin is easily removed, but emptying is messy, potentially exposing the user to dirt and allergens.

5. Vacuum cleaners traditionally move in straight lines, requiring the user to employ a series of push-pull motions to navigate corners and avoid obstacles, an inefficient and fatiguing activity. This problem is addressed by mounting the cleaner atop a large yellow ball, an innovation Dyson explored with his Ballbarrow, which enables the user to easily maneuver difficult angles with minimal effort. The pivot point sits between the body and the cleaning head, ensuring that the ball maintains contact with the floor at all times and keeps the motion fluid and uninterrupted.

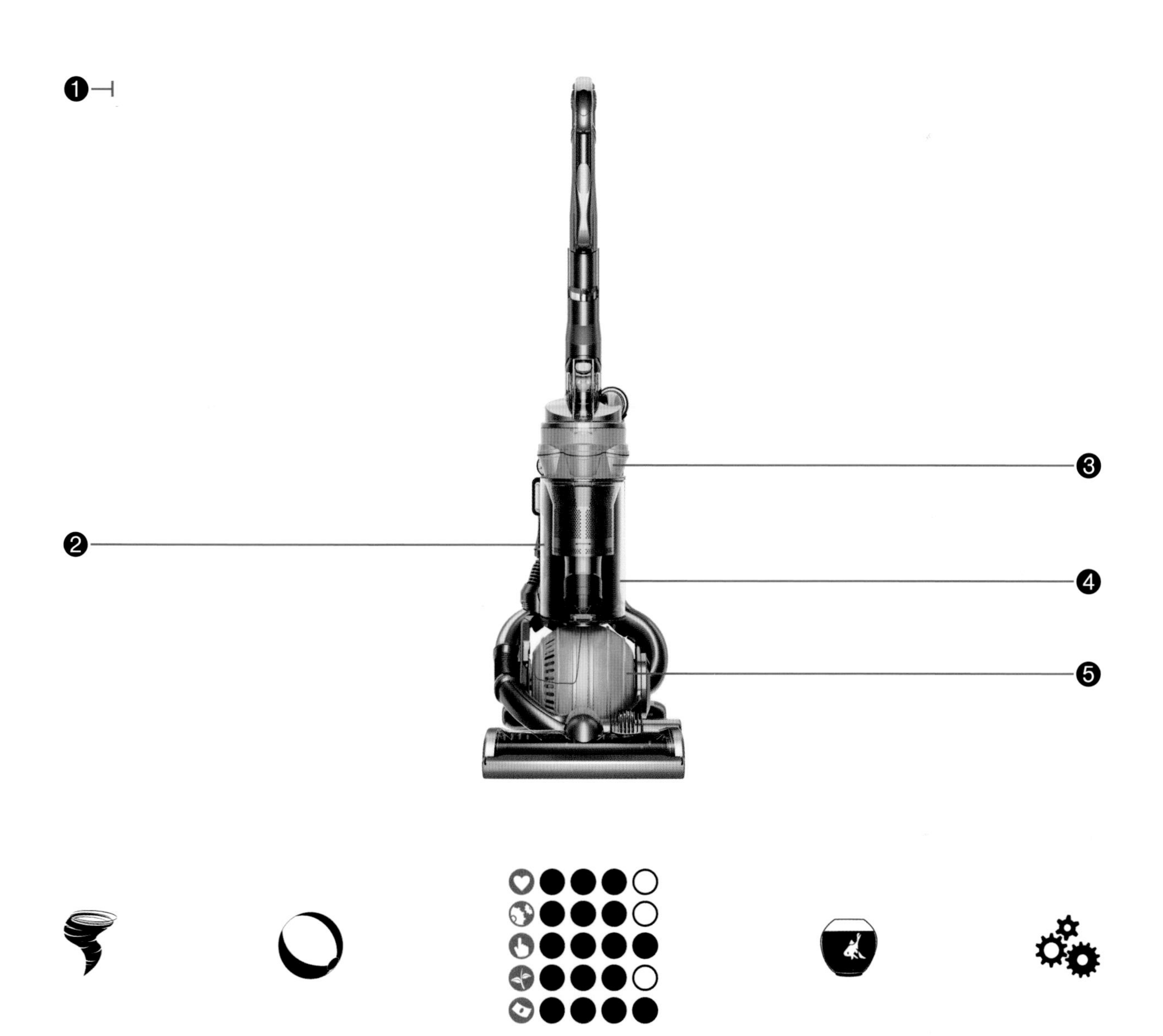
1
2
3
4
5

BeoCom 2 Phone

David Lewis and Bang & Olufsen, 2000

❶ As much sculptural piece as cordless phone, the Beocom 2 phone challenges virtually every phone design convention, from footprint to form to keypad. With a price approaching $1,000, the phone is not intended for everyone, and clearly targets a consumer willing to pay for design and status in their everyday items. Peter Eckhardt, managing director of Bang & Olufsen Telecom, comments: "With a design that is an open invitation to be picked up by a human hand, BeoCom 2 is proof that just because the telephone has become an everyday object, it doesn't have to be invisible."

❷ The body is a single tubular piece of aluminum formed into a gentle arc to conform to the face. The seamless metal body and simple form are distinctly modern, but the aluminum is cold to the touch, representing much of what is wrong with modernist design. The phone affords gripping and sits comfortably in the hand, but it is not easily cradled on the shoulder when two hands are needed. Acoustic performance is clear and sharp. The ring, created by Danish composer Kenneth Knudsen, is a unique and beautiful auditory expression of the phone, comprised of musical notes mixed with the sound of falling aluminum tubes.

❸ The keypad breaks with convention, utilizing a two-column layout that tapers to one that dovetails with the body. Although the keypad will offend the pattern memory of those who remember numbers by spatial position, the layout is easily mastered and affords a slenderness that allows the user to hold and thumb-dial the phone with the same hand. The button labels are painted on, like a common calculator, belying the high price of the phone. A pointing device is inserted low in the keypad, allowing navigation through a rich set of features such as an electronic phonebook and headset networking (for those with multiple phones). The display is a small backlit LCD panel near the top of the phone, which again reminds of inexpensive circa 1980s calculators and LCD watches. The compelling modernist statement that is Beocom 2 is at odds with its gadget-laden feature set, keypad pointer, painted labels, and liquid crystal display. Better to purify the design of its incoherence. Focus on quality, as with the acoustics and ringtone, not on quantity of low-value features. People are not buying this phone for its networking capabilities.

❹ The base is solid and heavy, giving the phone's high center of gravity a firm base while minimizing its footprint. The rounded base is beautiful, especially for a battery charger, and magnetized at the insertion point to facilitate docking. The docking interaction is oddly satisfying—the user need only get close to the docking port and the magnet assumes control of setting the phone to its proper home position. Once set, the phone and base resemble an exclamation mark, which seems an appropriate punctuation for the design statement.

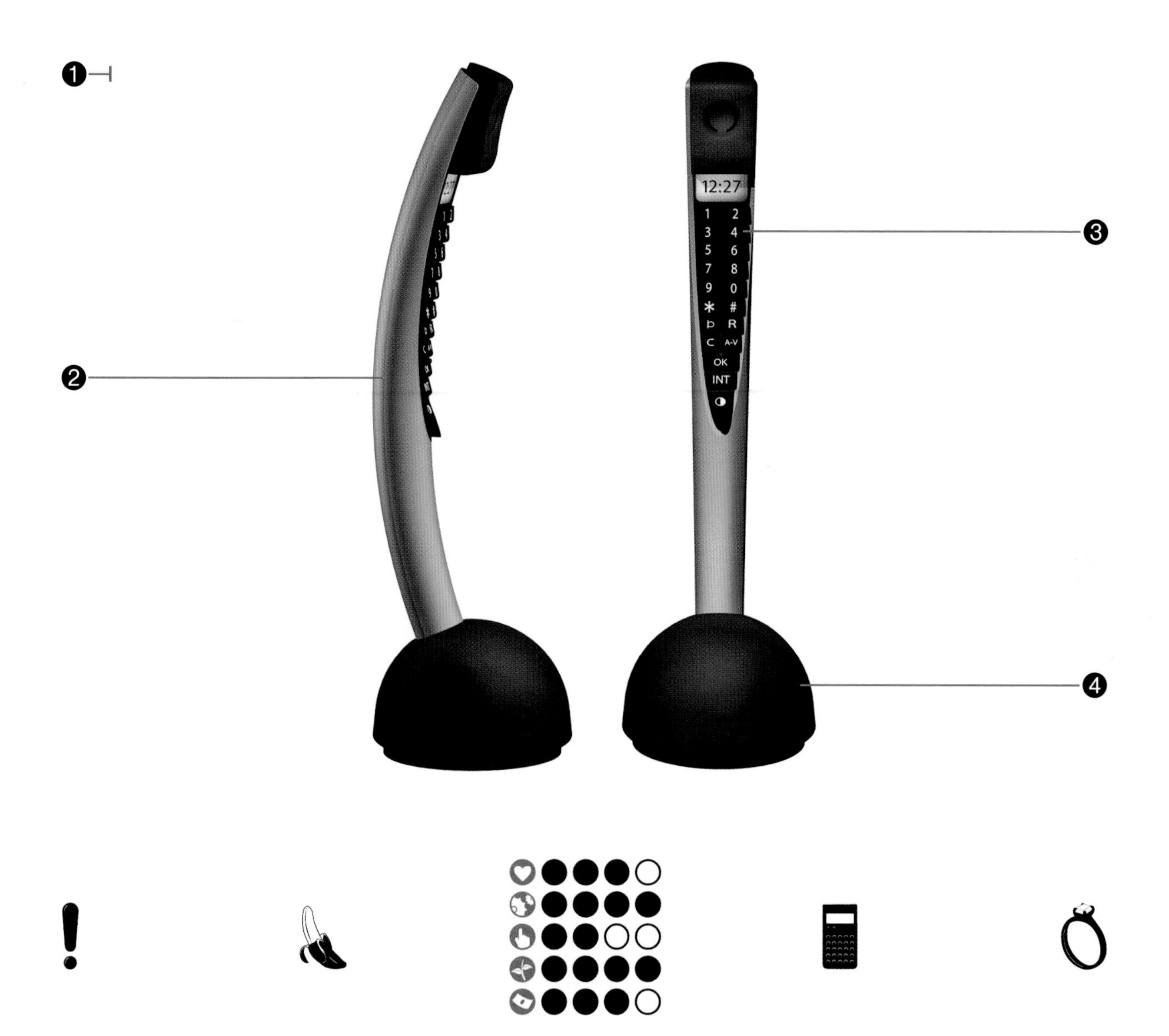
1
2
3
4
12:27
1 2
3 4
5 6
7 8
9 0
* #
R
C A-V
OK
INT

Bratz Doll

Carter Bryant, Isaac Larian, and Paula Treantafelles, 2001

❶ Prize to prepubescents and bane to parents, the Bratz doll has effectively supplanted Barbie as the aspirational image of Western adolescents. The dolls possess large, expressive faces, exotic names, multi-ethnic skin tones, and attire that could be kindly described as flirtatious. Why is it that tweens so favor Bratz dolls over Barbie? Paula Treantafelles comments: "At this age [seven- to twelve-year-olds] they're very different [than] four- to six-year-olds. They're about self-expression, self-identity. When Barbie was in her prime, girls were taught to be career women, to be men's equals. Today, yes, career and education matter, but it's also 'express yourself, have your own identity, girl power.' Strangely, Barbie might have missed that message."

❷ Increasing size by a factor of 7 to approximate human scale, Bratz would be 60" tall with measurements of 24"-15"-29." While this waist size is implausible at any stature, the key is the proportions. Low waist-to-hip ratios (WHR) are a reliable, cross-cultural predictor of overall female body attractiveness as rated by both males and females. Among normal populations, a WHR of 0.70 is considered ideal. However, innate biases such as this can be exploited—for example, waist training and corsets allowed women in the nineteenth century to achieve WHRs as low as 0.50. Bratz have a WHR of 0.52, a hyperfeminized proportion highlighted by their typically bare midriffs. By contrast, the chest-to-hip ratio (CHR) is 0.82, consistent with an adolescent body, giving Bratz a youthful appearance as compared with a modern Barbie who has a more mature CHR of 0.98. Bratz is what tween-age girls want to look like as teens. Barbie is what teenage girls want to look like as adults—the problem for Barbie is that most teenagers aren't playing with dolls.

❸ The head and face of the doll is the primary source of its appeal. The head is large and round, greatly out of proportion to the body, manga style. The eyes are oversized even relative to the enlarged head, the forehead is high, the nose and chin are virtually nonexistent—these superneonatal features are a clear, though likely unwitting, exploitation of the baby-face bias. Humans are biologically wired to find these features attractive. Additionally, the key facial features have been hyperfeminized and accented with makeup: red cheeks, full lips, and shiny hair are all signs of youth and reproductive fitness. Prepubescent females, in particular, are biologically wired to find these features attractive, as these qualities represent an archetype of fertility. It is important to note that whereas adults perceive hyperfeminized features as sexual, children do not—they are unconsciously attracted to such features as extreme signs of health, maturity, and reproductive fitness, not as sexualized objects. Why is it that tweens so favor Bratz over Barbie? It's not a "kids getting older younger" phenomenon. It's not "girl power." It's the face.

❹ The dolls are dressed provocatively but fashionably—trendy, skimpy, lots of bling. The style is more relevant to tweens than the more adultlike Barbie adornments, closely matching the style of popular teen idols and tapping the prepubescent's natural curiosity on how to achieve status, popularity, and ultimately mate attraction in the impending teen years. And the bling is not all vicarious: the Bratz Forever Diamondz comes with jewelry for its owner as well, complete with a real diamond and a certificate of authenticity.

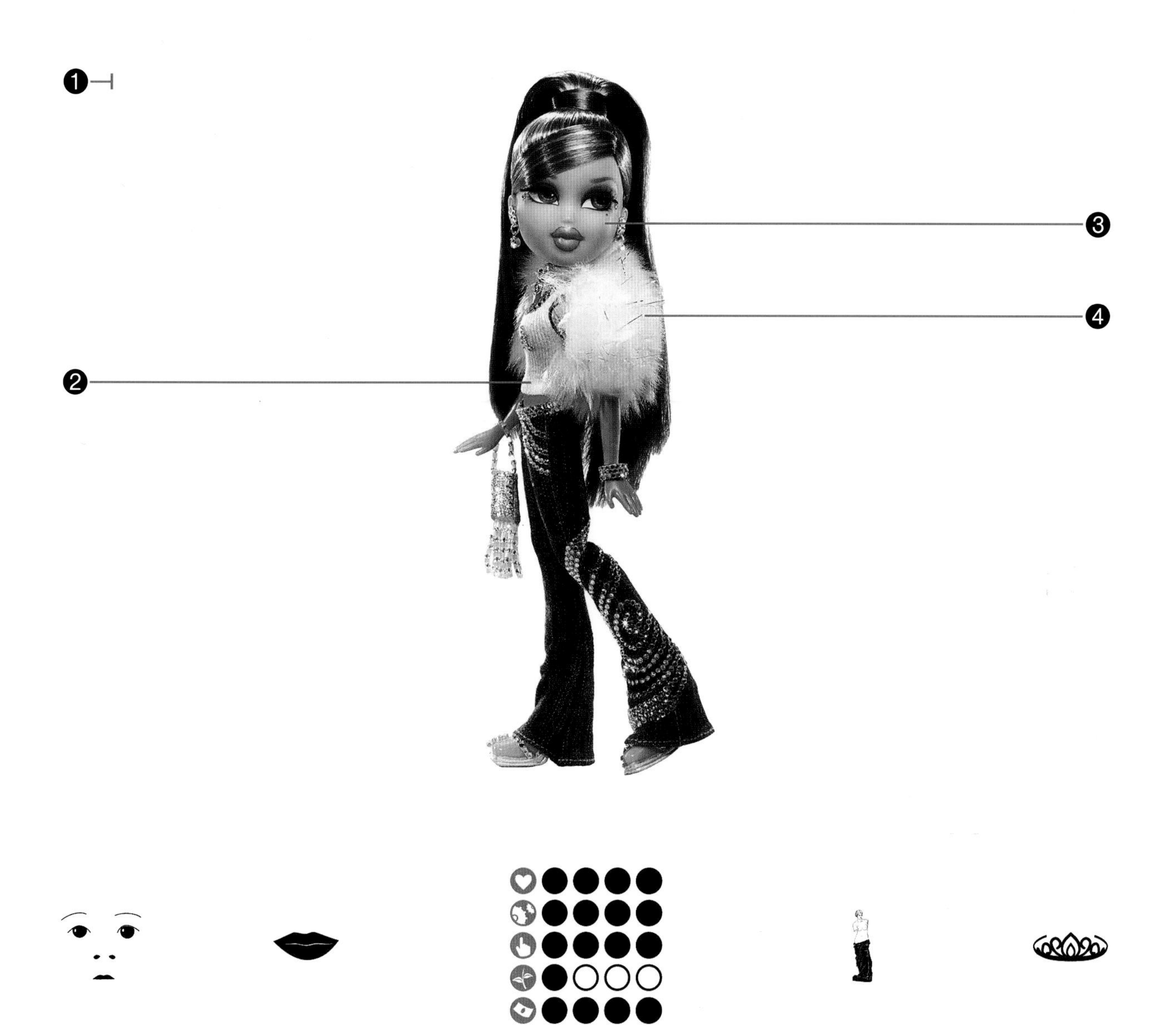

1
2
3
4

Bubble Lamp

George Nelson for Howard Miller, 1947

❶ Inspiration comes in many forms: a sight, a sound, a smell, and sometimes it is a must-have Swedish hanging lamp that is priced beyond your means. George Nelson comments: "It was important to me to have certain status symbols around, and one of the symbols was a spherical hanging lamp made in Sweden. It had a silk covering that was very difficult to make; they had to cut gores and sew them onto a wire frame. But I wanted one badly. We had a modest office and I felt that if I had one of those big hanging spheres from Sweden, it would show that I was really with it, a pillar of contemporary design. One day Bonniers, a Swedish import store in New York, announced a sale of these lamps. I rushed down with one of the guys in the office and found one shopworn sample with thumbmarks on it and a price of $125. It is hard to remember what $125 meant in the late forties ... I was furious and was stalking angrily down the stairs when suddenly an image popped into my mind, which seemed to have nothing to do with anything. It was a picture in the *New York Times* some weeks before, which showed Liberty ships being mothballed by having the decks covered with netting and then being sprayed with a self-webbing plastic ... Whammo! We rushed back to the office and made a roughly spherical frame; we called various places until we located the manufacturer of the spiderwebby spray. By the next night we had a plastic-covered lamp, and when you put a light in it, it glowed, and it did not cost $125."

❷ The Bubble lamps are widely acknowledged for their elegance and sophistication, though younger, unacquainted eyes could easily mistake them for inexpensive patio lighting. While the midcentury pricing for the lamps generally ranged between $15 and $45, downright affordable relative to the Scandinavian lamp Nelson pined for at Bonniers, modern replicas of his lamps range from hundreds to thousands of dollars. Given the source of its inspiration, one can only wonder what George Nelson would think of the fact that modern versions of his Bubble lamps have ironically become the variety of overpriced designer lanterns that made him so furious in the first place. The lamps came in a variety of bubble-like shapes and sizes. Nelson did not give these variations unique names, but they did receive unique model designations. For example, the "saucer lamp" was sold "Bubble Lamp H-727," the large "ball lamp" was sold as "Bubble Lamp H-725," and so on. The aesthetic of the lamps appear derivative — or an outright copy — of Klint's "Fruit Lanterns," which were popular in the period, though technically of Danish and not Swedish origin. In any event, Nelson's primary innovation regarded the adaptation of new material processes to inexpensively mimic and mass-produce a form that was perceived to confer value and status. Functionally, the lamps perform well, emitting diffuse, even illumination. The plastic shells are sturdy and durable, resisting damage better than their silk and paper predecessors, though they do tend to yellow with time and trap dust on the internal surfaces. A variety of accessories were offered with the lamps that could turn the "bubbles" into ceiling pulley fixtures, swag lamps, table lamps, and wall pin-ups, a flexibility that was rare for lighting of the time.

1
2

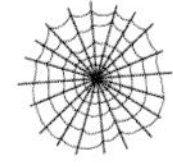

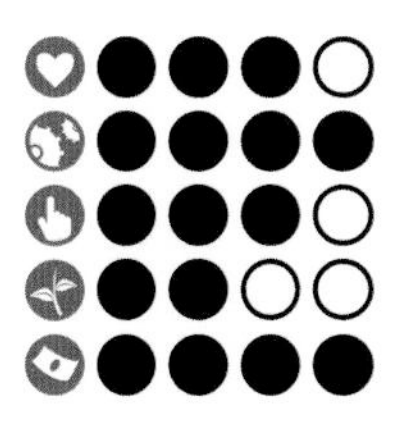

Cabbage Patch Kids

Xavier Roberts and Debbie Morehead, 1978

❶ No doubt most U.S. retailers remember the Christmas season of 1983 as "Invasion of the Cabbage Patch Kids," a line of dolls named after the 1901 novel *Mrs. Wiggs of the Cabbage Patch*, by Alice Caldwell. Although the dolls themselves appear harmless enough, the parental mobs that swarmed retailers to buy them were not—long lines and irrational exuberance combined with a limited supply of dolls creating a dangerous mix of hysteria and frustration that often resulted in violence and injury, all for want of a doll. What kind of doll drives otherwise normal people to this kind of behavior? Xavier Roberts comments: "The original idea was that they were really pieces of contemporary art, pieces of sculpture. In art school, that's what I wanted to do... I never saw them as toys. It was the adult that was always collecting them."

❷ At a time when the trend in doll design was mechanical motion, artificial pooping, and prerecorded recitations, Cabbage Patch Dolls represented a nostalgic return to simplicity, their arms permanently outstretched to be picked up and hugged. The dolls possess large, round vinyl heads and soft fabric bodies. The craft style is nostalgic, circa 1930s. Any appeal achieved through their baby-face features—large eyes, chubby cheeks, round faces, and short, pudgy limbs—is undermined by their oddly oversized mop of yarn-hair and hyperobese facial proportions. Every doll is one-of-a-kind in name and appearance, with variations of hairstyle and hair color, eye color, skin color, and clothing. This individuality is *bona fide* with a birth certificate signed by the father, none other than Xavier Roberts himself, and adoption papers conferring responsibility for love and care to the buyer.

❸ Given the questionable aesthetic basis for the popularity of the doll, the obvious candidates are their one-of-a-kind identities, elaborate backstories, and the ritual that accompanies their purchase. For example, never do the creators refer to the dolls as "dolls," but as "kids" born in BabyLand General Hospital (an actual medical clinic turned CPK tourist destination in Cleveland, Georgia, U.S.A.). One does not "buy" a Cabbage Patch Kid; they are "adopted" with the consent of their parents. When the box containing the CPK is first opened, prospective parents (i.e., buyers) are greeted with an unmistakable whiff of baby powder—what mere "doll" could smell like this? One does not "own" a CPK, but rather agrees to love and care for it. And, of course, "love and care" entails buying doll clothes and accessories, of which there is a wide selection. It is this grand story that creates the hysteria and mall mobs, not the doll. The story feeds the imagination, creates roles for all participants, and allows the buyer to be cast in the starring role of ultimate caregiver. The unique identities and pedigrees of the dolls allow adults to rationalize their purchases as investments in collectibles, but in truth it is the emotional, nurturing connection that drives their popularity, a connection intensified when additionally shared between a mother and daughter. This effective use of story and ritual by Cabbage Patch Kids has not been lost on competitors, most notably Build-A-Bear Workshop, where children actually build their teddy bears and other animals through an elaborate and highly interactive in-store ritual, and American Girl, where dolls have rich backstories from different historical periods that are recounted in accompanying books and in-store theater productions featuring the doll characters in a live-action musical.

2
3
Cabbage Patch Kids
Birth Certificate
Sherri Mazie
November 20th

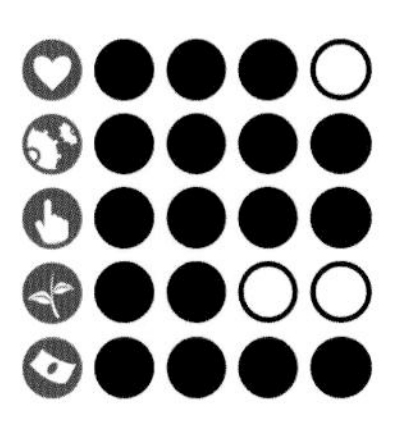

Cameleon Stroller

Max Barenbrug for Bugaboo, 2005

❶ The first Bugaboo stroller made its debut in the United States on the television program *Sex in the City*, a popular cable television series that all but defined big-city fashion in the early 2000s. Largely as a result, the high-end Bugaboo brand soon became associated with high-profile celebrity parents, including Madonna, Matt Damon, and Gwyneth Paltrow, to name a few, reinforcing its social function as a parental status symbol. Max Barenbrug comments: "Strollers used to be status symbols. If you look at the first strollers, they looked more like carriages, which is not strange because they were manufactured by carriage makers, and carriages also were status symbols. From the 1970s to the 1990s, strollers degraded. Stroller manufacturers decided to produce them in a cheaper way. We at Bugaboo came at the right time with our strollers; we made them worthwhile again."

❷ Of the various Bugaboo models, the Cameleon is the SUV of the line—stout, like a Hummer H2, with visible spring-loaded shock absorbers and rugged-looking, foam-filled, all-terrain tires. Bugaboo strollers are designed to appeal primarily to men, and the Cameleon is the most masculine of the bunch. The stroller supports innumerable adjustment and configuration options, including two-wheel or four-wheel mode (the former for sand and snow), reversible handlebars (small wheels forward for city, big wheels forward for rough terrain), and a reversible seat/bassinet. This flexibility does not come for free, however, and learning the various methods of flipping and folding the many sections requires commitment—the reality is that many users will find a comfortable general-purpose configuration and leave well enough alone. A wide variety of color combinations are available, all of which are bright and rich, akin to a sports car palette—no light pinks or baby blues allowed.

❸ Unlike traditional strollers, the Cameleon employs a single continuous handle instead of two independent handles—more like the handle of a push lawnmower versus the hook handles of an antique baby carriage. Not only does the single handle improve maneuverability, it enables stable, one-handed control while freeing the other hand. The handle also provides the structure necessary for conveniences such as a drink holder and related small-item storage, features noticeably absent from the otherwise bewildering array of stock accessories. These items can be purchased as *ad hoc* accessories, but it seems odd that a $900 stroller would lack a feature as basic as a drink holder.

❹ A disadvantage of being the Hummer of strollers is Hummer-like portability. The Cameleon is large and cumbersome for a stroller. It is not easily transported by car, and it is too large to fit in the overhead bin of a plane. The stroller can be collapsed and made more compact, but the effort required will lead many parents to simply not travel with it. The impressive flexibility of the many configurations sells well, but the resulting complexity exacts a toll in terms of usability.

❺ Perhaps the greatest design innovation of the Bugaboo strollers is the manner in which it exploits the parental need to validate, both to themselves and to those around them, that they are good parents—a need satisfied with an elixir of Veblen effect mixed with societal peer pressure. As sociologist Frank Furedi notes, "With lower-income groups, spending a lot of money is about saying 'we're respectable, caring parents.' With upper-income types, it's about saying 'This is how I am, this is a projection of me.' They don't even look at the price tag."

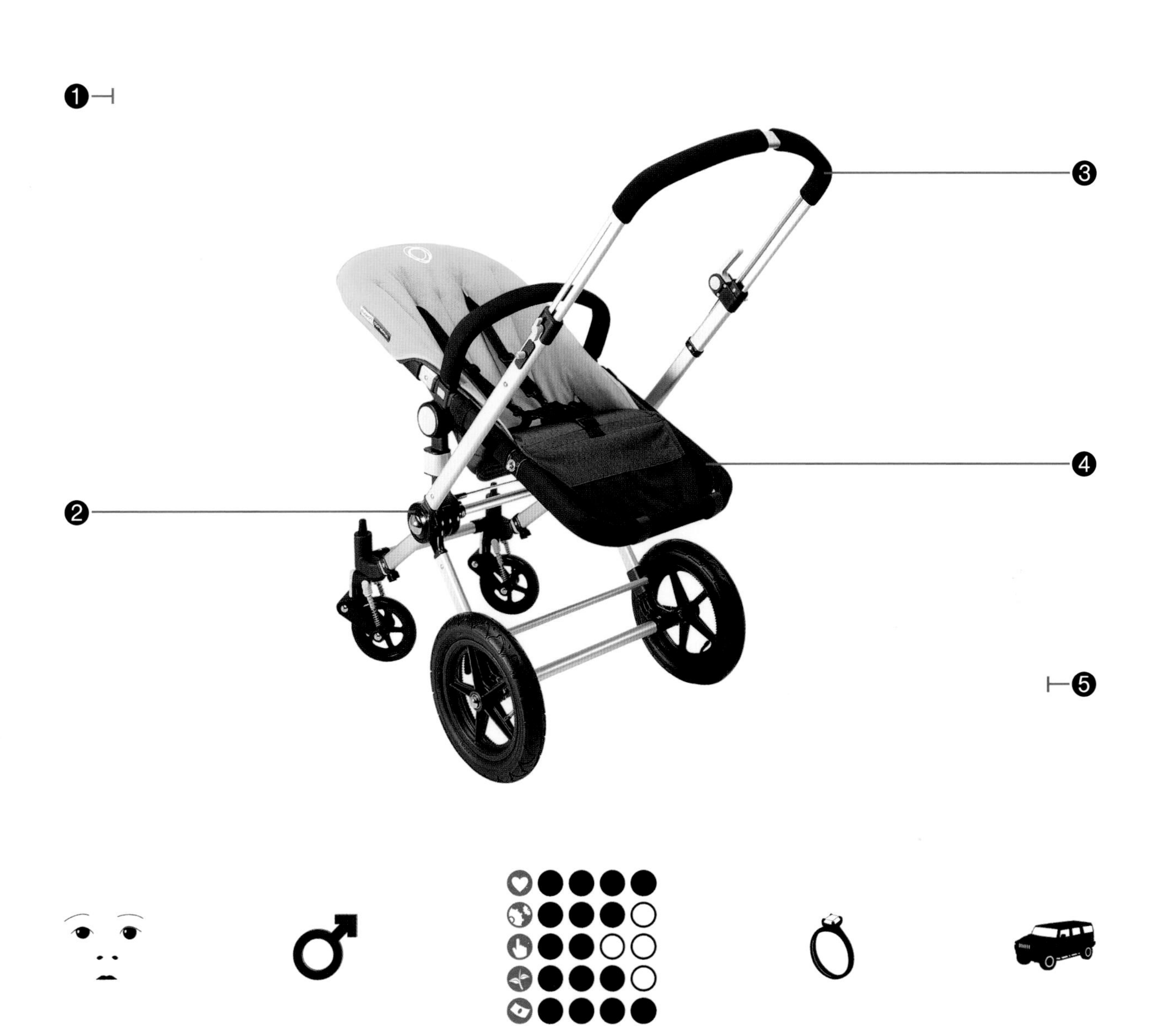
1
3
4
2
5

Cifra 3 Clock

Gino Valle, Massimo Vignelli, and Remigio Solari for Solari, 1965

❶ Imagine the perfect display for a digital clock. What would it look like? Would the type be segmented or jaggy? Would the legibility diminish at acute angles? Would contrast be a problem across lighting conditions? The answer to all of these questions is, of course, "No," but these are all common problems with digital clocks. It is amazing to think that in terms of legibility and overall usability, an analog clock introduced in 1965 with a split-flap display still outperforms virtually all-modern digital clocks.

❷ The clock is stark in its geometric purity, comprised of a cylindrical form with the front one-third scooped out for the display. Later models added a second smaller cylinder as the base support. The geometry is uninterrupted, however, as the cylindrical shape is maintained across a transparent plastic cover. The split-flap display is contained within a rectangular cutout with rounded corners, maintaining the radial theme. The cutout is left justified within the space and leaves considerable empty real estate to the right, perhaps a necessary accommodation of the internal mechanism, or a reserved space for the introduction of future functionality like a date display. Either way, matching the color of the display wall to the exterior—and resisting the need to fill the space with nonfunctional elements such as brand graffiti—ensures that the "white space" does no harm. Despite the many traditionally timeless elements of the design, the clock seems dated. Perhaps it is the large Helvetica typeface. Or the comparatively unrefined mechanical aspects of the flap rotation. As likely, a modern familiarity bias in favor of digital displays. Regardless of the cause, the effect is a clean, sophisticated product design that is aesthetically anchored to the era of its creation.

❸ The clock's split-flap display uses the same technology as the highly readable transportation departure boards found in airports and train stations. Unlike the segmented and illuminated characters of LED and LCD displays, the alphanumeric characters of the Cifra 3 are silkscreened onto plastic flaps in high-contrast white on black. In terms of legibility, the display is superior to even the high-end digital displays of today—type is cleaner and clearer and readable from acute angles. The transition between different states is, of course, more cumbersome than a digital display, requiring a mechanical rotation and flipping of the appropriate set of flaps. While this transition lacks the instantaneous update capability of a digital clock, this is an aesthetic quibble that certainly does not compromise time telling. In fact, for many, the experience of the flaps falling at precise intervals punctuates the marking of time, creating a moment of anticipation—an experience that served as the centerpiece of the movie *Groundhog Day*.

❹ Time setting is achieved using a semicircular control on the right wall of the clock, rightly anticipating that the majority of users approaching the clock will be right-handed. The control disk is discreet, but easily discoverable and usable. The disk has cutouts for fingertip pushing: one on top for setting minutes and one on the bottom for setting hours. The appropriate hour or minute flaps rotate forward when the control is nudged with the finger. This motion illustrates a minor limitation of the display technology: the clock cannot be wound back, as the flip mechanism operates only in one direction. So, come fall daylight saving time, the hour flaps must be rotated forward twenty-three hours to achieve the effect of winding back one hour.

2
8 30
3
4

City Hall Clock

Arne Jacobsen, 1955

❶ This classic clock was designed for the Rødovre City Hall in Copenhagen, Denmark, one of many accoutrements designed by Jacobsen specifically for the project. The clock has become an icon for functionalism. It distills the representation of time down to its barest essence without compromising readability. The clock's use of lines and circles to mark time is in fact more functional and usable than traditional numeric markers in that it is universal, enabling populations accustomed to Roman, Cyrillic, Arabic, Western, and other number systems to tell time with comparable efficiently without unfamiliar symbols or risk of alienation. Accordingly, the clock looks perfectly modern and at home anywhere in the world that follows base-12 time. Arne Jacobsen comments: "Functionalism is a style, or a liberation of a style, that expresses purely and clearly the function of things. Overstatement creates superficiality, indifference, and inauthenticity, which is precisely that from which functionalism flees."

❷ That the form of the clock looks so much like the office and administration lighting in Rødovre City Hall is no surprise; Jacobsens's fanaticism for continuity throughout his building projects is legendary. The face of the clock is simple and geometric, reminiscent of the compositional simplicity and rectilinear geometry of the building itself. Whether the clock appears bland or elegant depends one's taste for modernism. The face is covered with crystal glass and encased by an aluminum band. Hours are denoted by thick black lines, minutes by small black circles. The thicker hour hand falls well short of the hour marks, whereas the hour hand extends just short of the minute circles. The rounding of the hands and hour marks harmonize with the circular minute marks and circular face, visually connecting the elements and giving the clock strong overall coherence.

❸ There is perhaps no better example illustrating the oft-confused notions of functionalism and minimalism than the City Hall clock. Both concepts are rooted in modernism and are often used interchangeably, but they are not the same. A functionalist design can be minimalist, as with the City Hall clock, but it can also be relatively complex (e.g., PalmPilot). A minimalist design can be highly functional, but it can also be nominally functional (e.g., the Museum Watch). Functionalism regards meeting performance requirements. Minimalism regards achieving a pleasing aesthetic. Functionalism abides by an "everything follows function" philosophy. Minimalism abides by a "less is better" philosophy. Functionalism prioritizes function over other considerations, but does not dismiss or ignore other considerations unless they compromise function. Minimalism, by contrast, is an aesthetic philosophy, asserting that higher beauty is achieved by using the fewest elements possible in a design. It may very well dismiss or ignore other considerations in this pursuit. In a 1971 interview for *Politiken*, Jacobsen commented: "For many years it has been said that if a thing is practical and functional, then it is also beautiful. I don't believe in that at all, for there are different ways to work out a problem functionally, but beauty is a different matter." Jacobsen recognizes that functionalism is not a branch of aesthetics, does not directly speak to aesthetics, and is not an alternative to aesthetics. The two are independent and often competing factors. Many in the engineering community mistakenly embrace "minimalism" in the KISS (*Keep It Simple Stupid*) and Ockham's razor senses of the term, believing that simpler systems are typically more reliable than more complex systems, but this is really an appeal for efficiency in design, not an aesthetic reference.

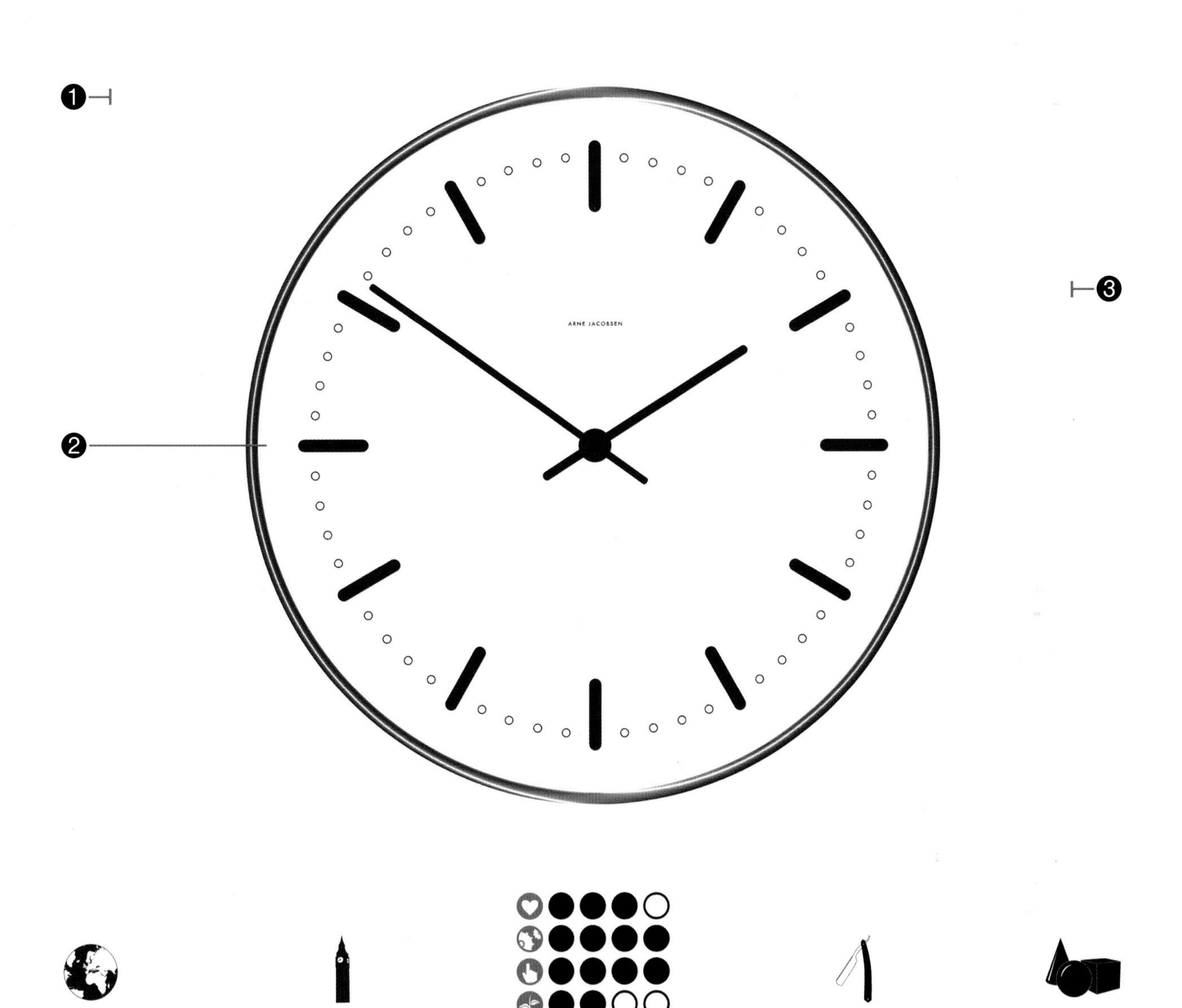
ARNE JACOBSEN
1
2
3

ClearRx Pill Bottle

Deborah Adler, Klaus Rosburg, and Target, 2005

❶ Traditional prescription pill bottles are rife with variability and clutter, which are key ingredients for human error. A poll by Target, for example, suggested that over 60 percent of patients had taken the wrong medication. Contrast this with the ClearRx pill bottle—simple, consistent, readable, and personalized—design priorities such as these are only achieved through an ardent commitment to user-centered design. Deborah Adler comments: “Our objective was to make three things clear: what the drug is, who it belongs to, and how to take it… This information needed to be communicated immediately… At the end of the day, the responsibility on how to take your medicine is yours. Most likely, the only form of communication you have is your medicine package, so it’s important that it is clear and easy to understand.”

❷ The bottle is rectangular with large, flat surfaces, permitting efficient information display without the wraparound problems intrinsic to round bottles. The bottle is meant to be stored cap down, enabling the three viewable surfaces—front surface, spine, and back surface—all used to convey critical information and printable using a single wraparound label. The front and back surfaces slope inward toward the spine, allowing the bottle to be easily picked up when packed tightly with other bottles on a shelf or in a drawer. Bump-outs in the plastic near the cap facilitate the patient’s grip when opening the container. The sides are uncovered to reveal the kind and quantity of medicine inside. The bottle is tinted red to convey caution and promote Target’s brand.

❸ The label observes a logical information hierarchy, aligning order of presentation, highlighting, and size with the importance of the information. The spine contains the name of the medicine in bold type for viewing from above. The front face presents first the name of the patient in small type, then the name of the medicine in bold type within a highlighted row that spans the width of the label, instructions follow in large, clear type, then detailed information about the prescription in smaller type, and lastly the Target name and logo at the bottom—this is the correct position for the name and logo, of course, but the guile required to get the marketing department to go on vacation the day of that decision has to be celebrated. The back face presents warnings with consistent icons each chunked within their own rows. The back label adhesive is open on one side, creating a pocket where additional information about the patient and medication can be included and physically attached to the bottle. It can also hold a card magnifying glass, which is available for free upon request.

❹ Cap rings of different colors are available to code the bottle to individuals, preventing households with multiple prescriptions for different family members from taking each other’s medicine. There is also an affective component to this: empowering family members to pick their own colors incorporates an element of personalization and “high touch” to the interaction, giving people who are sick some sense of control at a time when they are often feeling particularly powerless. Six colors are available.

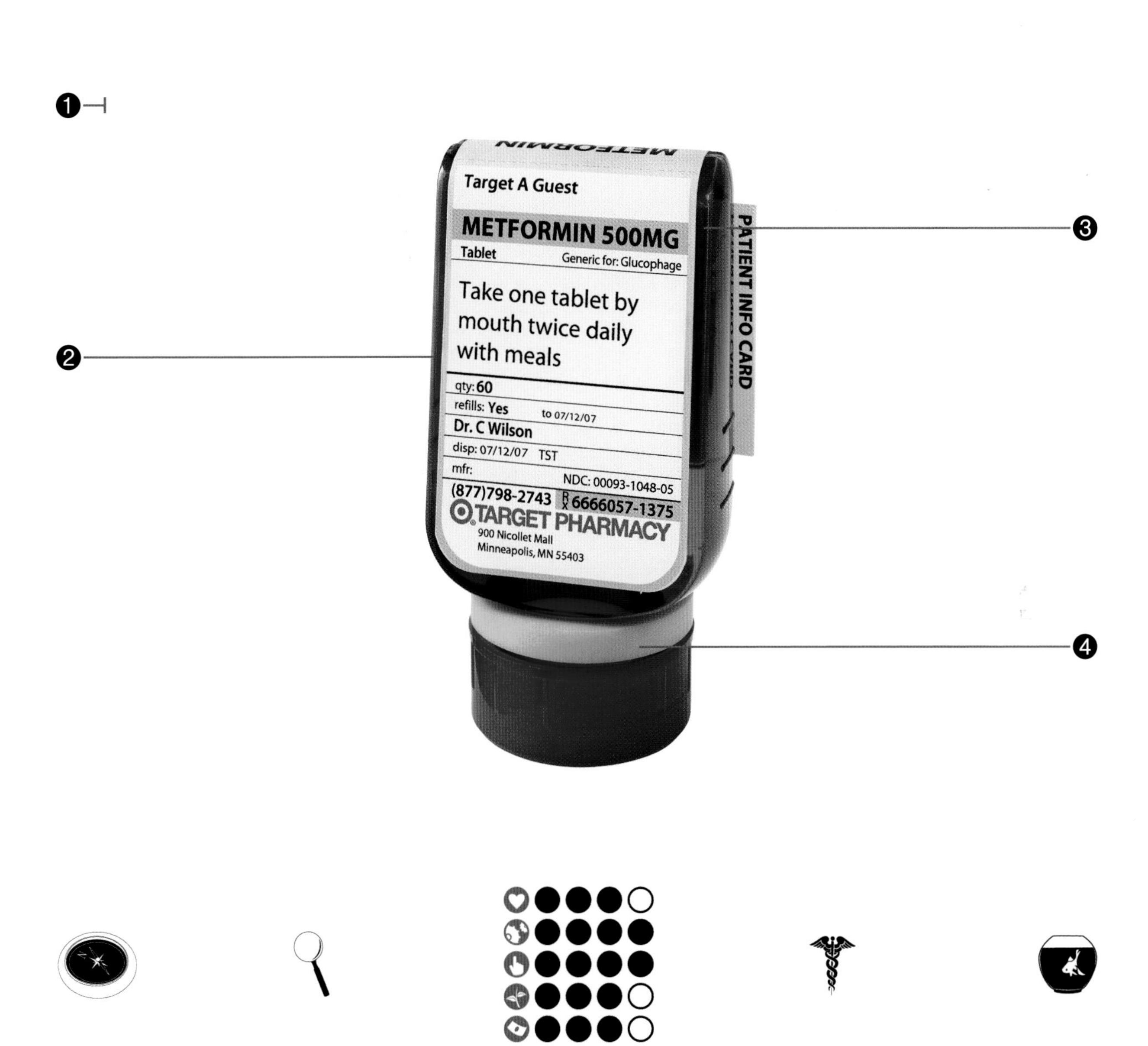

METFORMIN
Target A Guest
METFORMIN 500MG
Tablet
Generic for: Glucophage
Take one tablet by mouth twice daily with meals
qty: 60
refills: Yes
to 07/12/07
Dr. C Wilson
disp: 07/12/07 TST
mfr:
NDC: 00093-1048-05
(877)798-2743
Rx 6666057-1375
TARGET PHARMACY
900 Nicollet Mall
Minneapolis, MN 55403
PATIENT INFO CARD

Coca-Cola "Contour" Bottle

Earl Dean and Alexander Samuelson, 1915

❶ Frustrated by imitators in the early 1900s, the Coca-Cola Company launched a competition among its bottlers to develop a new bottle design that would distinguish it from competitors. The design brief read: "We need a new bottle — a distinctive package that will help us fight substitution ... we need a bottle which a person will recognize as a Coca-Cola bottle even if he feels it in the dark. The bottle should be shaped that, even if broken, a person could tell what it was." Seeking inspiration from the key ingredients of Coca-Cola at the time, the coca leaf and kola nut, designers from the Root Glass Company either mistakenly (or serendipitously) based their designs on a picture of the otherwise unrelated cocoa pod, an entry proximal to *coca* in the *Encyclopædia Britannica*. Early prototypes bore a resemblance to the female form, and were thus refined with due consideration to hobble skirt influences, a fashion fad of that period. The final curvaceous design won the competition and received a design patent in 1915.

❷ The form is distinctly feminine. Although the early bottles varied somewhat in their dimensions, the proportions approximate the aesthetic ideal for reproductively healthy females. At human scale, the bottle would have measurements of approximately 37"-24"-35," earning it the appropriate "Mae West" bottle moniker. Compared to the relatively bland and featureless bottles of the era, the curvy shape, soft fluting, and embossed logo gave the product a distinctively modern feel that stands to this day. The 1915 design has been tweaked over the years, but not bettered — the descendants of this timeless and soft-spoken bottle long ago joined the general grocery store din of insecure products posturing and shouting at consumers to get their attention.

❸ The logo is embossed, differentiating the bottle and making imitation more difficult. The embossment is applied to a smooth, unfluted middle region for reasons of manufacture. However, this interruption of texture combines with the verticality of the flutes to help the eyes find the logo — it appears to be center stage, featured between two sets of drawn curtains. The cost is confusion around the gripping affordance — is one to grip the lower narrow section of the bottle, or the middle nonfluted section? A quibble, as both areas are adequately grippable, but form should align with texture, clarifying, not confusing, its use. The typeface is Spencerian script, a standard style of penmanship practiced across the United States until the typewriter rendered it obsolete. The logo contrast is lacking by modern advertising standards, but is refreshingly subtle to more modern minimalist eyes. Make no mistake, however — the main objective of this bottle was to differentiate and protect the brand. An advertisement introducing the new bottle carried the headline, "We've Bottled Up the Pirates of Business." It goes on to tell consumers to "note the distinctive shape — the corrugations — the name Coca-Cola blown into the bottle. Fix the picture in your minds. It is sure protection against imitations and substitutes. In the future, accept no bottled beverage as genuine Coca-Cola unless it comes in this bottle."

❹ The narrow waist is naturally grippable, and synergizes with the raised fluting, helping to create an interesting tactile experience, what Coca-Cola referred to as an "in-hand embellishment." The glass is heavy, solid, the weight of the glass disguising the modest six-ounce volume of the bottle — a disguise further enhanced by the fluting, which accentuates and magnifies the cola contents.

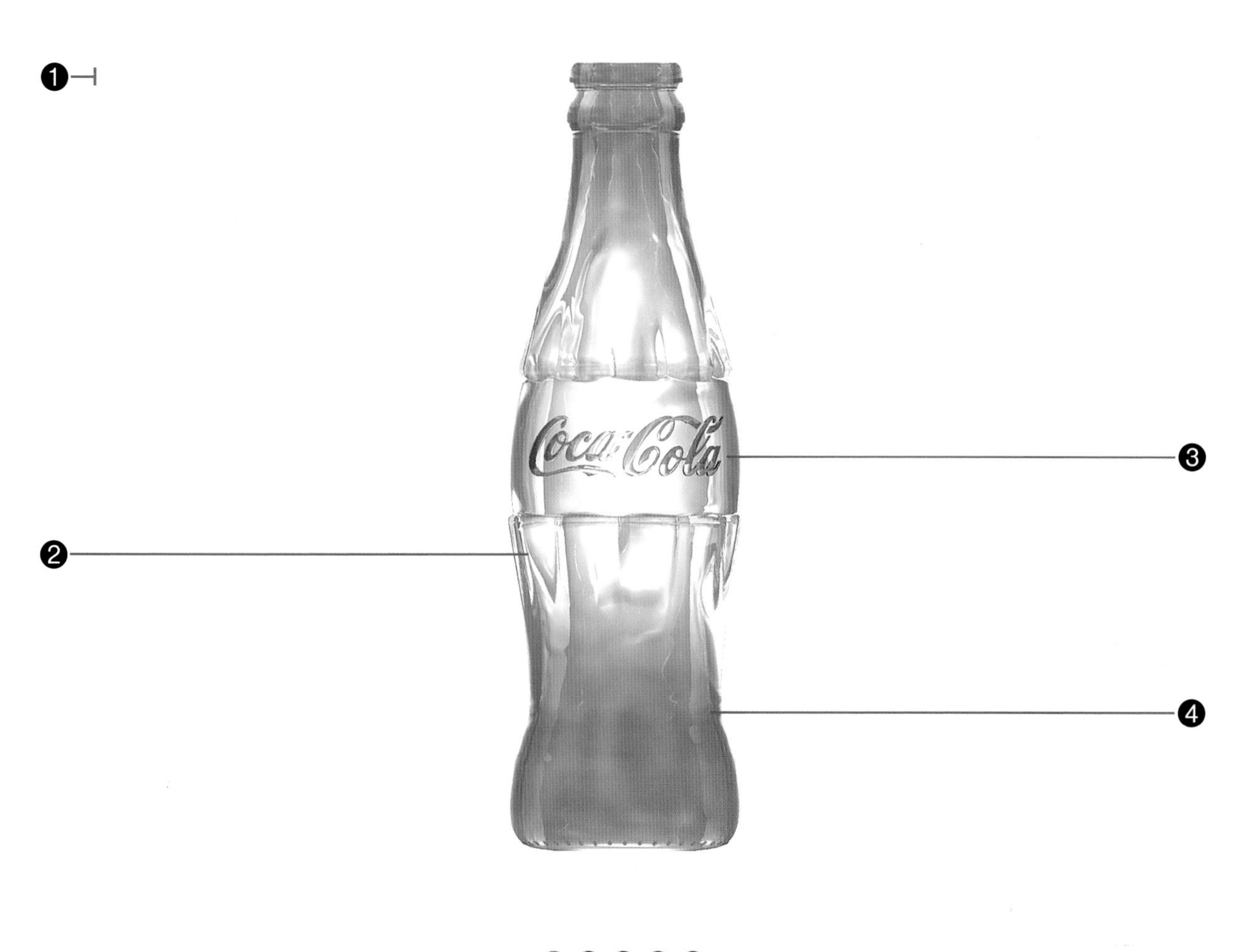
1
2
Coca-Cola
3
4

Coldspot Refrigerator

Raymond Loewy for Sears, 1935

❶ In what many consider to be the product that launched "industrial design" as a profession, the Loewy-redesigned Sears Coldspot refrigerator demonstrated that industrial design, done right, could have a swift and significant impact on the bottom line, more than tripling refrigerator sales for Sears in one year. Raymond Loewy comments: "When we started our design, the Coldspot until then on the market was ugly. It was an ill-proportioned squarish box 'decorated' with a maze of moldings, panels, and other schmaltz. It was perched on spindly legs high off the ground and the latch was a pitiful piece of cheap hardware. We took care of all that in no time at all. The open space under the box was incorporated into the design and it became a storage compartment. The new latch was substantial and as well designed as if it had been intended to be the door handle of an expensive automobile, the hinges were made unobtrusive, and the name plate looked like a fine piece of jewelry. The entire connotation was one of high quality and simplicity."

❷ The long, conspicuous handle affords pulling, which in this case runs counter to its function. The handle, described as the "feather touch" latch, was designed so that a housewife could easily open the door with her hands full of groceries, using her elbow to depress the handle. Its action is that of a big button, which gently releases the door latch when depressed. The design of the handle is counterintuitive—that is, affords pulling when pushing is required—but the solution is an early example of how better understanding the usability requirements of a product can lead to innovations that have high problem-recognition appeal to consumers, which directly impacts sales.

❸ Prior to Loewy's redesign, the Coldspot appeared to be modeled after period furniture, along the lines of a Chinese wedding cabinet. While the redesign introduced a number of innovations like nonrusting aluminum shelves and an easily cleaned polish finish, most significant was creating a new and unique identity for refrigerators—a cross between a vault and a stylish, streamlined automobile. The form is clean and white with rounded corners and rolled edges. One large door with one long push handle replaced the more complex French-like doors of previous models. The resulting product looks stout and high quality—like a safe place to store food.

❹ If the tall door with its centered vertical molding looks like something from the hood of a streamlined midcentury car, it should come as no surprise that much of the inspiration for this refrigerator came from Lowey's work in automotive design. The idea for aluminum shelves, for example, was inspired by his work designing hood grilles. The basic form of the refrigerator was one of the first products to be prototyped and modeled in clay before being pressed into metal, again a common practice in automotive design but new to product design. In addition to making the refrigerator appear interesting and reminiscent of cars, the vertical molding made the product appear taller, implying greater volume. The logo plate is prominently located near the top centerline, text on polished metal, not unlike a large hood ornament.

❺ The Coldspot was one of the first products to plan design refinements on an annual release cycle, an attempt to motivate consumers to replace their current refrigerators based on fresh design and compelling new features.

❶
COLDSPOT
❸
❷
❹
❺

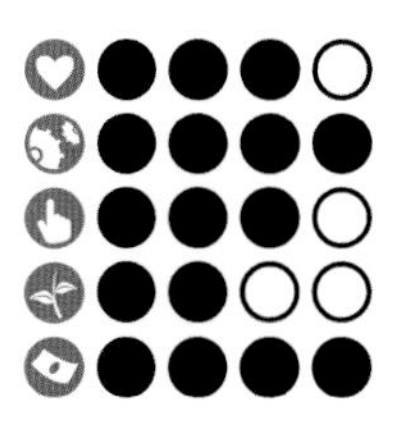

Crescent Adjustable Wrench

Karl Peterson and Crescent Tool Company. 1908

❶ Contrary to popular belief, the name "Crescent wrench" did not come from the tool's resemblance to a crescent moon, but from the name of the company that produced it, the Crescent Tool Company. Inspiration for the adjustable wrench came when a visitor described an ingenious adjustable wrench he had seen in Sweden. Karl Peterson quickly developed a wooden prototype of the tool based on the description, and a steel version soon followed—the Crescent wrench was born. The tool was received amazingly well by consumers, and quickly became the must-have tool for makeshift mechanics and diehard adventurers. Charles Lindbergh flew across the Atlantic with only "gasoline, sandwiches, a bottle of water, and a Crescent wrench and pliers." Richard Byrd took Crescent wrenches with him on his South Pole adventures. And more than one NASA space mission took Crescent wrenches along in case of Apollo 13–type emergencies. Karl Peterson comments: "We did a lot of impossible things in those days, but since none of us had ever been to college, we didn't know they couldn't be done."

❷ Its form is remarkably organic for its time, a departure from the tradition of rectilinear tools. Although coincidental, the crescent-shaped head and jaw served the brand well, visually reinforcing both the tool and company name. The wrench is easily interpreted: the head looks like a traditional wrench head, but with a movable lower jaw. The toothed rack in combination with the worm gear immediately indicates how the wrench can be adjusted, especially to the more mechanically minded consumers of that period. The worm gear is the only actionable element, an affordance supported by its knurled surface. And, of course, no tool would be complete without a hole in the handle for hanging.

❸ The adjustable wrench is roughly equivalent to nine fixed wrenches. This amazing flexibility, however, comes at a cost, described as the flexibility-optimality tradeoff. This principle asserts an intrinsic performance tradeoff between specialized versus flexible designs—that is, gains in flexibility necessarily require sacrifices in performance relative to a more specialized designs, and vice versa. This tradeoff for the Crescent is demonstrated in mechanical slop resulting from the elasticity and imperfections in the sliding jaw's bearing. The consequence is a tendency to deviate slightly from an adjusted width, especially under load, often slipping and rounding the edges of the bolt or nut. Whether the cost benefit is justified depends on the situation. In situations where optimality is key, as in a professional garage, fixed wrenches will generally be preferred. In situations where flexibility is key, as with a field repair, flexibility will generally be preferred. A functional measure of a design is the degree to which a product delivers on one end of this continuum without compromising the other. By this standard, the Crescent wrench delivers in spades.

❹ The wrench was highly imitated, but the precision fabrication methods developed by Crescent Tool were both patented and difficult to imitate. The tool developed a reputation for superior quality, a reputation bolstered by the company's policy to replace any tool returned by a customer for any reason. It is testament to the performance of the design that a Crescent wrench today is essentially unchanged from the one introduced over one hundred years ago.

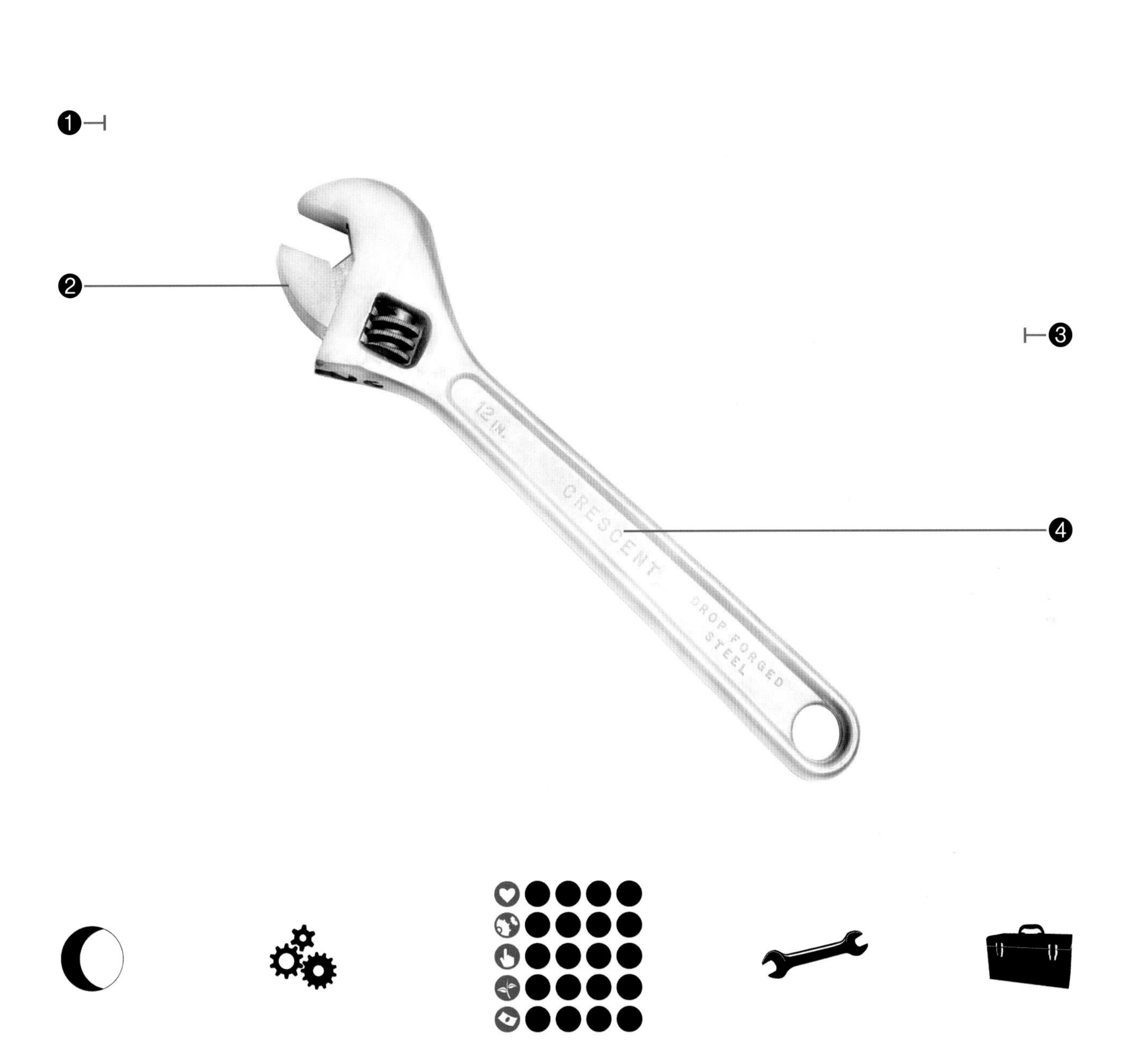

1
2
3
4
12 IN.
CRESCENT
DROP FORGED
STEEL

Cristal Ballpoint Pen

Décolletage Plastique Design Team for BIC, 1950

❶ Invented by Hungarian László Bíró in 1938, the ballpoint pen began life as an expensive and unreliable alternative to the fountain pen—more of a novelty item than a practical writing implement. It was not until a chance encounter with a wheelbarrow that Marcel Bich would find the inspiration to transform this prototypical pen into the ubiquitous writing implement that we know today. Bruno Bich, son of Marcel Bich and chairman and CEO of Société Bic, comments: "My father told me that one day he was pushing a wheelbarrow when it dawned on him that the ball was a multifaceted wheel and this was the best way to convey ink. So he put all his investment into the ballpoint." In 1950, Marcel Bich bought the patent from Bíró, refined the design and manufacturing processes, and gave the product brand his last name, sans the *h* to avoid any unfortunate Western mispronunciations. Since its introduction, the Bic Cristal has become the iconic expression of a ballpoint pen, boasting lifetime sales exceeding 100 billion pens worldwide.

❷ The pen was designed to be simple, reliable, and cheap. Given its success and how little the pen has changed since its introduction, it is clear that its overall design is near optimum for its category. The hexagonal shape of the barrel promotes gripping and prevents rolling, though can be hard on fingers with extended use. The barrel is constructed of clear plastic, revealing the pen's interior design and enabling the user to instantly see the amount of ink remaining. As Bruno Bich notes, "The idea was that there should be nothing superfluous and you could see how it works and how much ink is left."

❸ The cap is removable and color-coded to indicate the color of the ink. Modern versions bear a small hole in the top to reduce the choking hazard for children. The pocket clip initially made a statement about the manufacturer's confidence that the pen would not leak, and perhaps as a result it makes the pen ride quite high in the pocket—too high. Additionally, the caps get lost or chewed on, and the pocket clip is more often used as a hygienic pick of some sort than for its intended purpose. After more than fifty years, the pen is ripe for a redesign that loses the cap.

❹ The ball in the ballpoint is formed from tungsten carbide powder, vitrified with additives to diamond-like hardness to ensure that it will not pit or develop flats on its surface. The ball resides inside a brass socket that is joined to the ink tube. The fit of the ball is tight enough to prevent air from contacting ink in the ink tube, but loose enough to allow the ink to stick to the ball as it is rolled, which is the basic mechanism for applying ink to paper. A small hole in the middle of the barrel allows air into the barrel so the rotating ball creates only a partial vacuum, ensuring steady ink flow and minimizing leaks under pressure (e.g., in airplanes). Alternatives exist, but the simplicity and performance of this configuration have proven difficult to better. And contrary to urban legend, a million-dollar pen is not needed to write in spaceships or space stations—the Cristal ballpoint pen will work just fine in zero gravity.

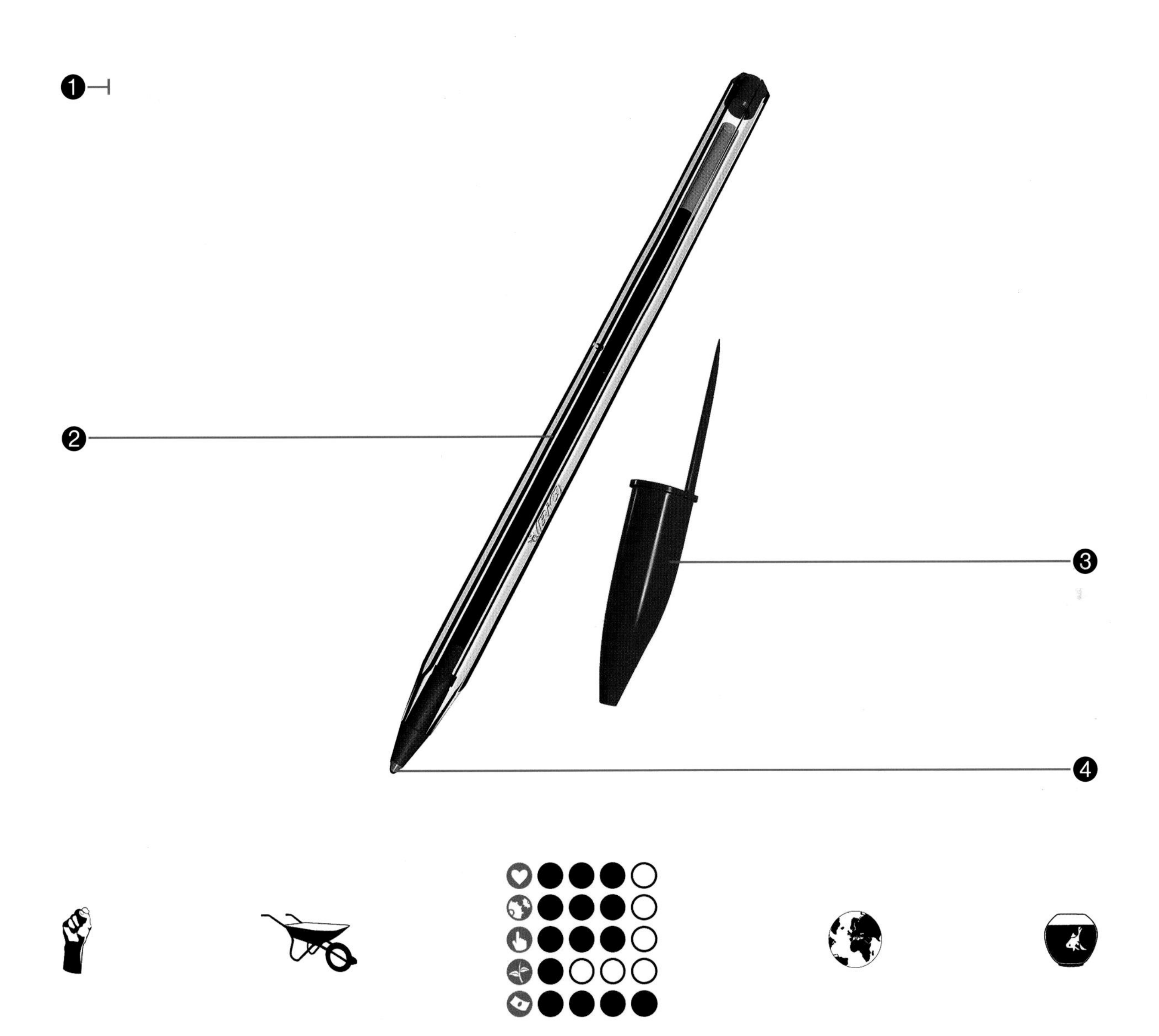
1
2
3
4

Crocs Shoes

Ettore Battiston and Crocs, 2002

❶ Reviled as homely and acclaimed as comfortable, Crocs shoes play in what can be called the comfort wear movement—a movement that puts comfort first and foremost, and fashion second, if at all. Crocs began its life as a humble boat shoe, but word spread fast about the remarkable comfort of the odd-looking clogs. Demand grew, and distribution quickly moved from boating conventions to small retail outlets to Nordstrom and Dillard's. Crocs became ubiquitous, achieving global footwear phenomenon status in less than two years. But what goes up must eventually come down, and as of this writing Crocs (the company) is facing several challenges: invalidated patents, market saturation, and a fashionista backlash, all resulting in a drop in stock price and sales. Can Crocs regain their phenom status? The answer hinges on whether management comes to realize that, contrary to appearances, their competitive advantage never really had anything to do with shoes.

❷ The heel strap rotates and stretches, enabling easy foot entry and securing the shoe to the heel when walking, differentiating the shoe from common clogs and avoiding the "flop" of flip-flops. The strap is hinged to the shoe with a pin that bears the Crocs logo, a cartoonish crocodile that, when coupled with the bright uniform colors of the shoes, emanates a decidedly childish quality—an idiosyncrasy that threatens the duration of the Crocs' fad cycle. Teen and adult models would be well served to either change the logo or lose it altogether.

❸ The shoe is roundy, cloglike, and uniform in color, presenting a friendly effect. It makes feet appear larger than they are, but its large size minimizes interior contact with the foot, making it a viable option for those with foot sores or circulation problems. The holes and applied surface textures appear reptilian, visually reinforcing the brand name. The principal innovation in the shoe design regards its material, a proprietary closed-cell resin called *croslite*, which is waterproof, odor-resistant, antimicrobial, nonmarking, comfortable, lightweight, recyclable, and warms and softens with body heat—one wonders why the manufacturer limits the application of this amazing material to shoes.

❹ Aside from Crocs' croslite construction, their most interesting element is their holes. Originally designed to drain water and promote air circulation, the holes inadvertently enabled a secondary market: the creation of decorative buttons, most popularly known as *Jibbitz*, that snap into the holes, enabling the wearer to personalize and decorate the shoe. Not only does this help address the wanting aesthetic of Crocs, it lets the wearers do the addressing, empowering them to remedy the key design flaw of the shoes they so otherwise love: to make beautiful what was once ugly.

❺ The shoes contain an orthotic, molded foot bed and built-in arch support for comfort. The insoles of the shoes are lined with nubs to provide traction, stimulate circulation, and displace accumulated moisture. This is a subtle but interesting feature. Had the design used dimples versus nubs, a plausible design, foot stimulation and water displacement would have been sacrificed. The difference between product success and failure is often as small as a 1 mm nub versus a 1 mm dimple.

2
3
4
5

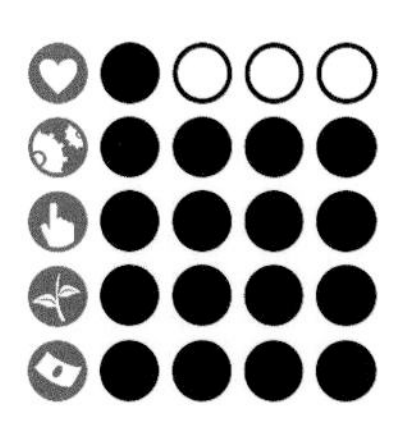

Diamond Chair

Harry Bertoia for Knoll, 1952

❶ The Diamond chair, one chair in a collection of wire-frame chairs, was created in a converted studio-barn under a Medici-like arrangement with Knoll, which gave Harry Bertoia complete artistic freedom to explore art and sculpture with the hope that the some of the fruits of his labor would have commercial potential. The investment paid off. The wire-frame line of furniture earned Knoll international acclaim and ensured that Bertoia could continue creating (and eating). Harry Bertoia comments: "In furniture, as in sculpture, I am concerned primarily with space, form, and the characteristics of metal. In the chairs that I had designed for Hans and Florence Knoll, many functional problems had to be satisfied first. I wanted the chairs to fit as comfortably as a good coat. For flexibility in the basket construction, I used steel for strength and unified the design with nickel-chrome plating for beauty, giving the chair durability and lightness in appearance ... If you will look at these chairs, you will find that they are mostly made of air, just like the sculpture. Space passes right through them."

❷ Like Gehry's Wiggle chair, the Diamond is more sculpture than chair, an exploration of space at the intersection of geometric and organic design. The chair scores in terms of originality and visual interest, appearing to contemporary designer eyes like a physical manifestation of an unrendered 3D model. However, in terms of usability, it is simply not comfortable. Seat cushions and stretch cover pads are available, which largely ameliorate issues of comfort, but at the cost of the chair's dramatic aesthetic. Bertoia often commented that the Diamond was made mostly of "air"—an apt description of how it looks, but not of how it feels.

❸ The Diamond chair—its name, shape, transparency, and chrome finish aligned to support both a literal and metaphorical expression of the chair's preciousness and value—invites sitting. The spreading of the grid approaching the seat acknowledges gravity, and draws the eyes and body to the seat. The expanding grid also implies an elasticity that is contradicted by no less than the tensile strength of steel. The design plays with expectations in this way, and its unique aesthetic buys it much forgiveness for those who do not find the irony amusing. The bucket is ample and gently curved, accommodating a wide variety of body types. The armrests are long and wide, and extend outward from the center as natural and uninterrupted extensions of the form. The front edge of the seat dips slightly as it should, but the impact on the lower legs is unmitigated. Sitting for more than a few minutes at a time requires a cushion.

❹ The leg assembly appears *ad hoc*, as if Bertoia lost interest once the challenging spatial problems of the Diamond seat had been solved. The goal was obviously to make the seat appear to be hanging, like a hammock or net draped between points of suspension, but the implementation looks more like poorly constructed lawn furniture. Early models attached the seat by using rubber shock mounts, which allowed users to rock back and forth. The dissonance between the elegantly filigreed seat and its clothes-hanger-like legs is visually grating, rivaling the leading contender of design disharmony, the LC4 chaise and its hyperbanal base. The legs are stable when unperturbed, but a high back combines with an inadequate rear leg angle to make the chair susceptible to tipping. Not a child-friendly chair.

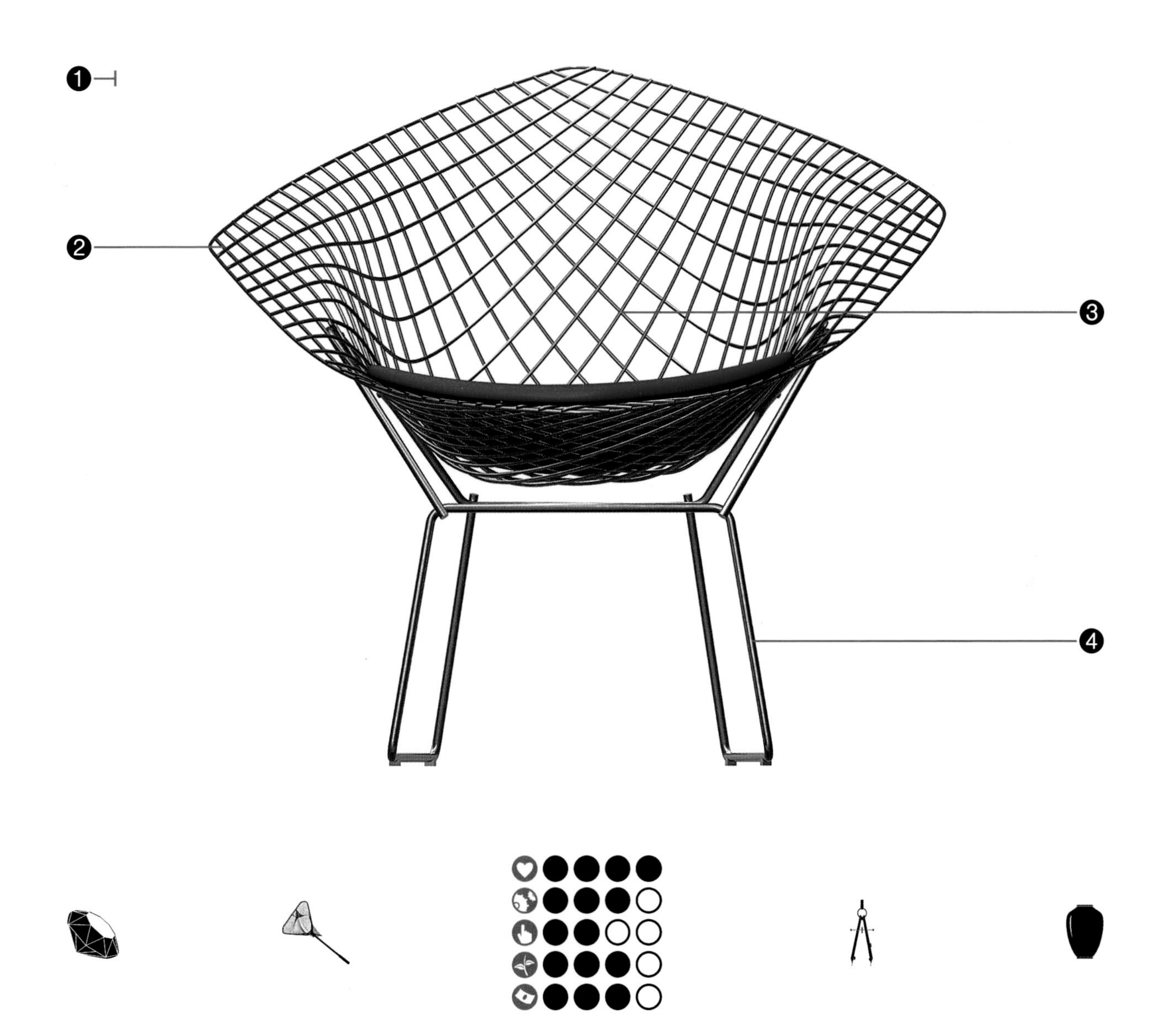

❶
❷
❸
❹

Dish Soap Bottle

Karim Rashid for Method, 2002

❶ Hidden beneath a sink in most kitchens is a bottle of dish soap—except, that is, for those kitchens with the Method dish soap bottle designed by Karim Rashid, where the dish soap sits proudly atop the sink to be seen and appreciated. The design stems from what Rashid calls *democratic design*, a philosophy based on the belief that design should be available to the masses. And where better to apply this philosophy than the kitchen? Karim Rashid comments: "It's the project I [had] been waiting for all my life… It was a chance to put a piece of sculpture on every American's sink."

❷ The form is organic and bloblike, exemplifying a greater design movement that will no doubt one day be referred to as *blobjectism*. Steven and Mara Holt Skov assert in their book, *Blobjects and Beyond*, that the popularity of blobject design indicates that people desire objects that tell stories and establish anthropomorphic and emotional connections, aspirations certainly achieved with this design—the wide base, narrow neck, and large, round head invites anthropomorphic interpretation, and provide the basis for the nickname, the *dish butler*. Much of the emotional connection here, however, is more baby-face bias than butler, as the disproportionately large and bulbous head appeals to our innate bias for baby-shaped features and proportions. Aesthetically, the bottle easily doubles as modern sculpture. It redefines what a dish bottle can be, humanizing and modernizing an otherwise unnoticed iconic form. Functionally, the narrow neck affords gripping, while the large head prevents the bottle from slipping through soapy hands. Dispensing soap is intuitive, but hindered by a rigid plastic body that resists squeezing a bit too fervently.

❸ The bottle is injection-molded translucent polyethylene. The surface is lightly textured, improving grippability and hiding the use factors that can make it look dirty (e.g., fingerprints). The frosted-glass appearance reinforces the appearance of modern glass sculpture and also gently reveals the type and quantity of the soap remaining. The tactile experience is interesting, soft, and organic, but the rigidity of the plastic is otherwise, and requires some squeeze strength to expel soap. Given the compressive resistance at the neck, the user must strangle this poor anthropomorphic form to get soap out, and invariably control over the amount of soap released gets lost in the struggle.

❹ The logo and type are simple and modern, employing understated lowercase letters cast in white to contrast with vibrant soap colors. Messaging is painted onto the bottle, avoiding the inevitable wear and peeling of traditional labels as they get wet and degrade. Again, the bottle maintains the appearance of "clean" throughout its use cycle. Placement of messaging is below the neck, leaving the large head unmarked, which reinforces anthropomorphic interpretation.

❺ The bottle rests with the dispensing valve at the bottom, eliminating the need to invert it to use it. A narrow, rapid soap stream combined with a wide base minimizes soap accumulation around the valve, and effectively hides what does accumulate. Placing the valve at the bottom, however, places a greater functional burden on the valve's ability to seal—and perhaps predictably the valves are prone to leakage, especially in the early models. A small (and technically correctable) deficiency given the many positive attributes of the design.

1
2
3
4
5
m
method
dish soap
ultra concentrated formula
aroma: lavender
25 FL OZ (1.56 PT) 739mL

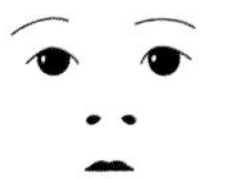

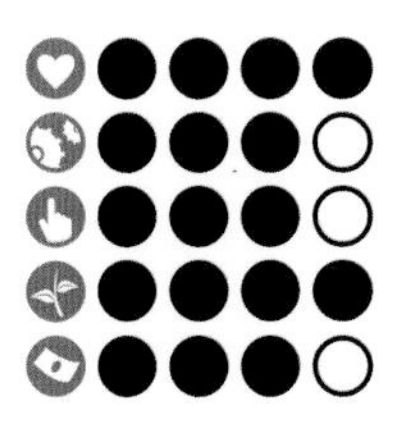

Dixon Ticonderoga Pencil

Joseph Dixon Crucible Company, 1913

❶ Affectionately known to most as the "yellow No. 2 pencil," the Dixon Ticonderoga was first introduced to American schools in 1872, and has since become the iconic pencil. Joseph Dixon did not invent the pencil, but he invented pencil automation, helping make a dry, clean, and portable high-quality writing implement available to the masses. Although it may be easy to look past pencils today, to take them for granted, it is important to see the pencil through a historical lens. Then, and only then, can one see and appreciate the product as the Dixon Ticonderoga Company clearly does, as more than a mere writing instrument, but as an indispensable recorder of thoughts and feelings. The company's description reads: "Dixon Ticonderoga is a company that empowers people to take conscious and subliminal thoughts, facts, ideas, and dreams, and preserve them using tools that are simply extensions of themselves."

❷ The body of the pencil is hexagonal in cross section. The simplest shape to manufacture is a square cross section, but this form is uncomfortable to the hand and difficult to sharpen. The most ergonomic shape for writing is a triangular cross section, which provides large flats for each of the three gripping fingers, but it, too, is difficult to sharpen. The preferred shape for drawing is a circular cross section, which enables easy adjustment and gripping from a variety of angles, and it is optimal for sharpening. In terms of material use, however, square, triangular, and circular cross sections use more wood than a hexagonal cross section, making them more expensive to produce. For reasons of economy of materials and as a compromise between the various usability requirements, the hexagonal cross section prevailed in the marketplace and remains the dominant standard today.

❸ The addition of rubber erasers to pencils occurred in the mid-1800s, effectively representing the first integrated undo-like functionality for writing on paper. Their successful integration, however, was not universally accepted until the early part of the twentieth century, at which time nine out of ten pencils sold were eraser-tipped. According to an essay in the Dixon catalog weighing the pros and cons of "plain vs. rubber-tipped pencils," the arguments against eraser-tipped pencils included increased cost, poor-quality materials and manufacture, reduction in student performance by promoting errors, and disease transmission (from chewing).

❹ Dixon was one of the first manufacturers to use circular lead rods—the early standard was square rods, which were easier to manufacture but weaker and difficult to sharpen. Hardness is a function of the amount of clay mixed with the graphite, and is indicated by numbers on the body (1 = extra soft to 4 = extra hard).

❺ In the late 1800s, Franz von Hardtmuth, grandson of the founder of Faber pencils, resolved to make a premium pencil that would sell for three times the price of common pencils. Since the graphite used was deep black, he decided to paint the pencil yellow, in honor of the black and yellow Austro-Hungarian flag. The current ubiquity of yellow pencils, however, has little to do with flags. Consumers superficially associated the yellow body of the pencil with eastern Asia, which was believed to be source of the finest grades of graphite. Yellow symbolized quality, and the pencil, named Koh-I-Noor after the famous diamond, became a great success. Other pencil manufacturers quickly learned the lesson and began painting their pencils yellow. The color became the *de facto* standard for pencils, and to this day three out of four pencils sold are yellow.

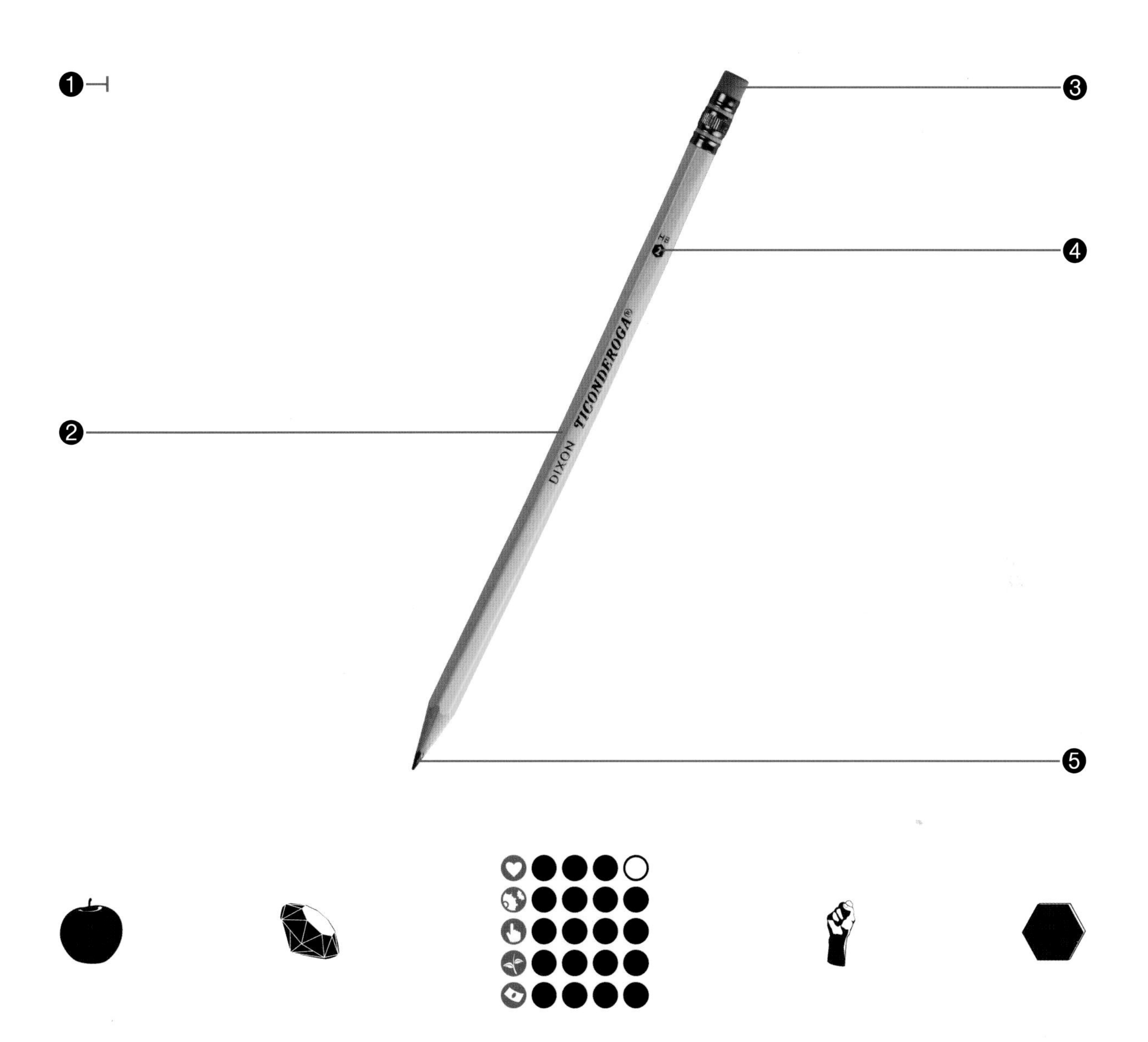
DIXON TICONDEROGA®
HB
2

Egg Bird Feeder

Jim Schatz, 2004

❶ The Egg Bird Feeder enhances backyard vistas, attracts a diversity of small birds, and resists both hungry squirrels and contrary weather. Jim Schatz comments: "The design followed the egg, which is the basis of life. The Egg Bird Feeder is designed to be beautiful and functional—an object that complements the beauty of wild birds and deters squirrels from disturbing them." Squirrels, however, are resourceful critters, and can, in many cases, get to the booty—but the extraordinary acrobatics involved will more than compensate the owner for the transgression.

❷ The body is hollow and made of ceramic earthenware; a continuous enclosure except for the connecting holes at its apex, which nominally expose its contents to rain runoff from the suspension cables. Its slippery, glazed surface and ovoid shape resist both the elements and leaping squirrels. The wide middle impedes visibility and access to the food tray from the top, while the narrowing bottom creates a hollow that permits easy access from the perch rods. Seed is transferred from the interior of the body to the feed tray through force of gravity. The outer lip of the feed tray rises in height just above the exit of the body, preventing overflow. The feed tray has four drain holes to prevent water accumulation. The design exhibits remarkable coherence—name to form, metaphor to material, symbolism to function—with an economy of elements. The materials are simple and natural. The palette is limited to two colors: the spun aluminum finish of the suspension cable, perch rods, and feed tray, and the single lively color of the ceramic body. Each feeder is handcrafted with modernist precision, and is proof that form need not suffer for function.

❸ The feeder is suspended at two points by a slender coated cable, which disappears nicely against complex backgrounds. This two-point suspension is stable when light loads (e.g., small birds) are applied to the perch rods, but unstable when heavier loads (e.g., squirrels) are applied. The cable also serves as a barrier to the center axis, reducing its affordance as a jump destination but increasing its gripping affordance.

❹ Two aluminum rods penetrate the lower body at right angles, creating four equidistant perches. The perch design keeps approaching birds from colliding and allows different species to eat from the feeder at the same time. Holes in the lower walls of the body support the perch rods, which in turn support the feed tray. A single hitch pin connects the perch rods from below and fixes them in the center position. Filling the body with seed is achieved by inverting the feeder, removing the hitch pin, perch rods, and feed tray, and then filling. This is no small affair. It means that the feeder must be taken down and moved to a location where it can be safely disassembled and restocked. Disassembly requires at least two hands, and the fragile egg does not naturally sit inverted waiting for seed—this is where the price for its round, smooth, featureless surface is exacted, because the shell is easily dropped and will roll off tables if not secured. The practical consequence of all of this is that only diehard birdwatchers will properly maintain and restock their Eggs. Everybody else will use the feeder a couple of times, then reconsider whether they really want a high-maintenance designer feeder or a zero-maintenance designer yard ornament—and most will opt for the latter.

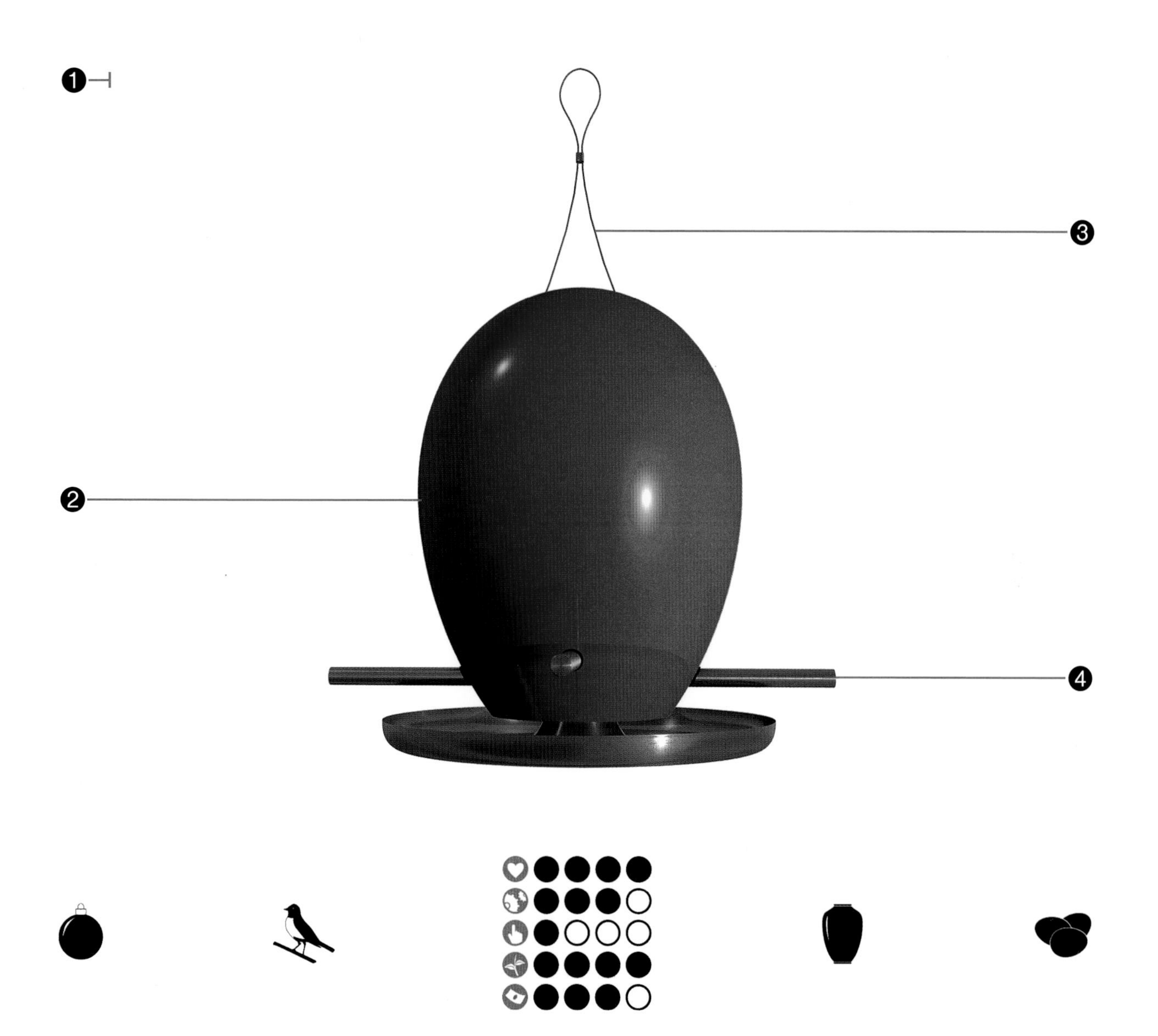
1
2
3
4

Forehead Thermometer

Scott Henderson for Vicks, 2008

❶ Consider the traditional mercury-in-glass thermometer. It is made of glass, takes an indeterminate amount of time to achieve a final reading, is difficult to read, and contains the highly toxic element mercury. Does this sound like a viable product? An instrument that should be inserted into a child's body? These questions underscore the tendency of people to habituate to existing product norms, however unacceptable, until a new product comes along that highlights the deficiencies. The Forehead Thermometer is such a product. It not only addresses the many problems of mercury thermometers, it addresses the many problems of modern conductive thermometers—and does so in manner that pays special attention to the needs of children. Scott Henderson comments: "The key to the design of the Vicks Forehead Thermometer was to balance the ideas of medical credibility with a nonthreatening, juvenile aesthetic because it's primarily a product for young children. The magnifying glass was a perfect metaphor to achieve this, as a magnifying glass is both scientific while simultaneously embodying 'youthful exploration' or 'new discovery.' The large display also helps the parent see a reading at full arm's length, contributing to the noninvasive nature of the product."

❷ The form resembles a toy magnifying glass—simple and unthreatening. The thermometer measures temperature by analyzing the infrared energy (i.e., heat) radiated from the skin above the eyebrow. Temperature is taken by placing the thermometer flush against the forehead just above one eyebrow, pressing the power button, and then slowly moving the thermometer laterally back and forth until the device beeps. A temperature reading takes approximately three seconds. While similar devices have a reputation for inaccuracy, the test unit performed comparably to a digital oral thermometer.

❸ The display is large and its LCD display clear. In addition to a numeric reading, the background lights in one of three colors indicating the degree of health risk associated with the temperature: "acceptable" is coded green, "elevated" is coded yellow, and "high fever" is coded red. This not only translates the meaning of the temperature in practical terms, it accommodates temperature taking in dim light, which is often the case with children. Instructional icons and simple animations direct user actions, but their implementation suggests a lack of user testing. For example, a brow-swipe animation is initiated when the unit is first activated, appearing to indicate that the device is ready and waiting for the user to apply and swipe the thermometer across a sick person's brow. However, the device is actually waiting for the power button to be pressed again, an action required to actually take the temperature. The result of this ambiguity is users believing they are doing what they are told, following the on-screen prompt and swiping the device, while the device does nothing waiting for the power button to initiate the measurement. Despite the clumsiness of the software interface, users will likely figure it out eventually—the benefit of having only one button. Still, an unfortunate mar on an otherwise intuitive design.

❹ The thermometer has one primary button; it is convex and marked with an icon that appears to symbolize radiated heat. A more discretely positioned side button stores twelve temperatures in memory. The thermometer has all of the benefits and liabilities of being based on a single multifunction button. There is no doubt how to activate it, but the function switching involved in its operation is confusing, and can lead to frustration.

1

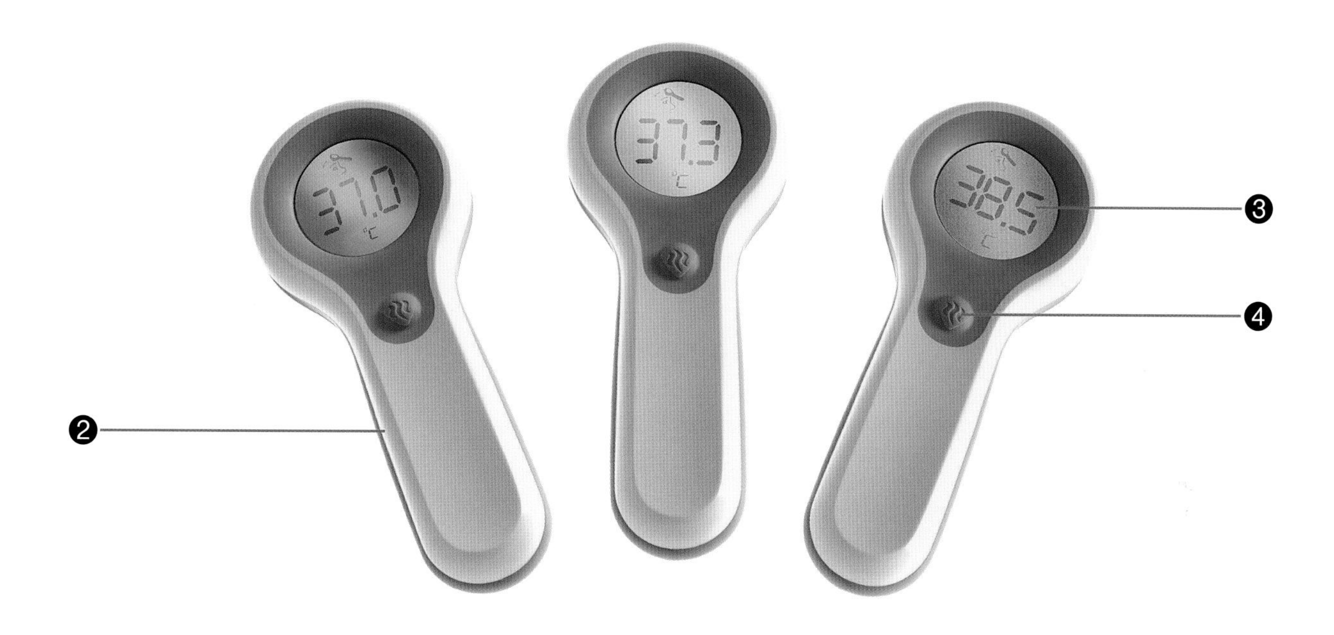
37.0
°C
37.3
°C
38.5
°C
2
3
4

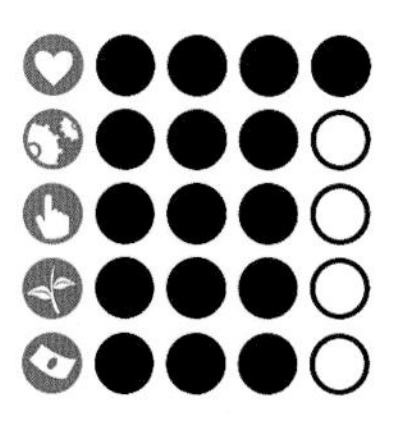

FuBar Functional Utility Bar

David Richardson and Stanley Tools, 2006

❶ It is said that when your only tool is a hammer, every problem looks like a nail. The genius of the FuBar is the recognition that this age-old maxim is no longer true for modern construction workers. For this user group, when your only tool is a hammer, every problem looks like something to demolish—those now customarily armed with nail guns use their hammers for demolition. The challenge then becomes actualizing this new understanding in the redesign of a product that hasn't significantly changed in hundreds of years. Joe Martone, Stanley product designer, comments, "We wanted to make it look aggressive, almost medieval, so that nobody would mistake it for an ordinary hammer."

❷ The tool is large and foreboding, drawn from a single piece of forged steel and tempered to resist chipping. If the objective was to make it look medieval, mission accomplished—it looks like destruction incarnate. The name FuBar is a double entendre, an abbreviation of *Functional Utility Bar*, and, of course, the more colloquial interpretation from WWII, *Fucked Up Beyond All Repair*. Given the target audience and the design's focus on demolition, the name couldn't be cleverer. Unfortunately, this cleverness is lost in the marketing-speak of its full name: Stanley® FatMax® Xtreme™ FuBar® Utility Bar.

❸ Serrated jaws effectively grab and twist dimensional lumber and tear drywall. The jaw span is two-tiered for lumber with 1" and 2" widths, the most common in housing construction. The upper tier juts outward to a thin wedge, useful for both penetrating drywall and prying tight joints. The triangular stamp arguably reinforces the lower jaw, but given the lack of editing elsewhere, it is more likely superfluous ornament.

❹ The striking surface combines a fairly traditional hammer head with a long, wedged neck, the latter of which is particularly useful for breaking and shearing. The functional beauty of the form is undeniable, but as with its name, the manufacturer's *horror vacui* could not leave well enough alone: "XTREME" (no "E" and, of course, all uppercase) is prominently stamped into the flat of the neck—for those who might mistake the design as subtle.

❺ The handle is long enough for two hands when needed, and the rubber grip is easily managed dry or wet. The top of the grip is cut at angle to continue the line of the lower jaw. The lower grip sports subtle cuts on its edges and a checkered texture on its sides to enhance gripping. The company logo, legal messaging, and colors are appropriately represented on the handle.

❻ The bottom of the tool serves as a rough-nail pull and crowbar, and the teardrop cutout as a fine-nail pull. The crowbar-end curves away from the holding arm of right-handed users, but curves inward toward the holding arm for the left-handed, making one-handed striking for lefties impracticable. The length of the tool earns its stripes when prying and pulling nails, providing ample leverage aided by a laterally large and imminently grippable head, which supports pulling, pushing, and twisting.

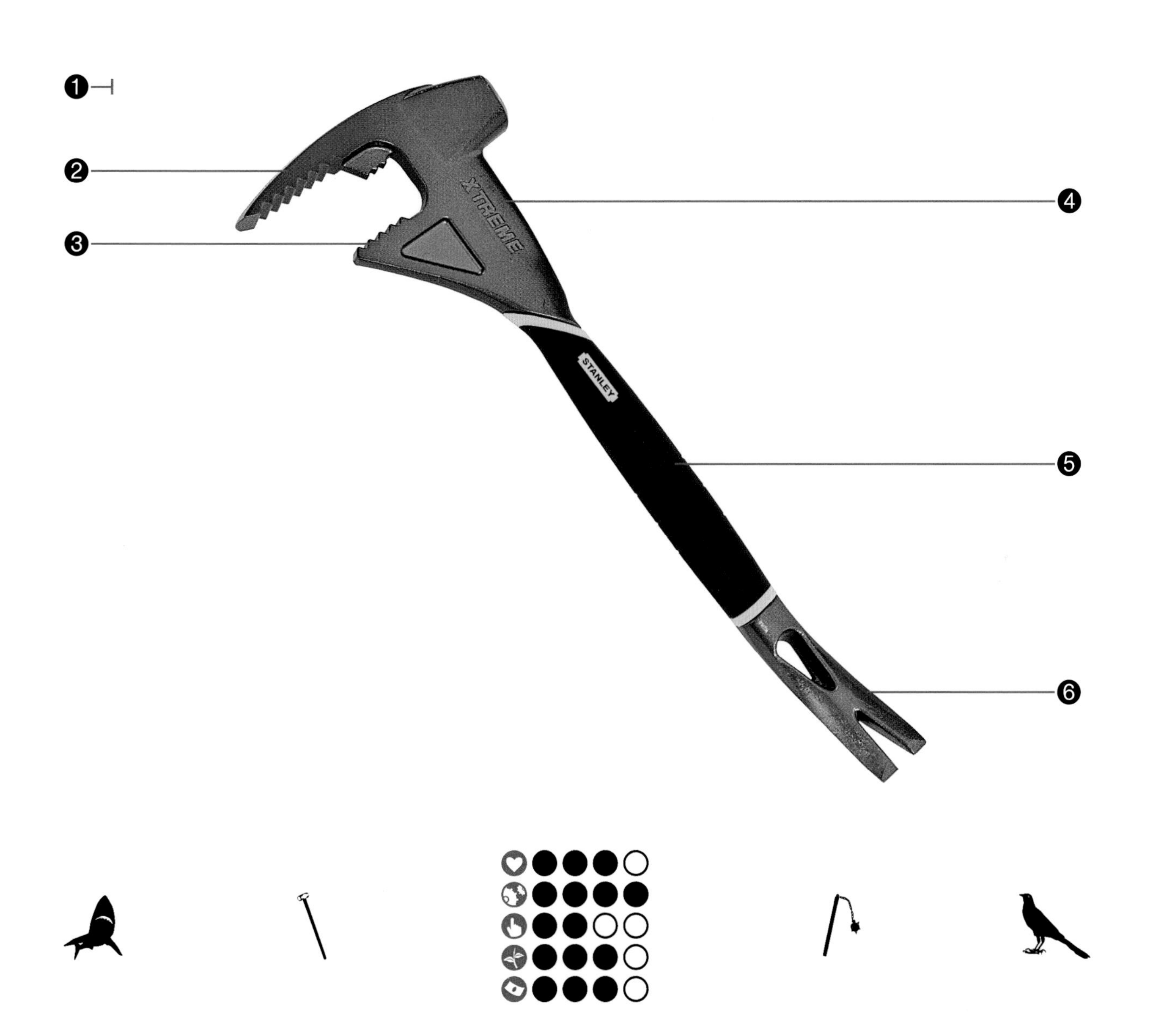
1
2
3
4
5
6
XTREME
STANLEY

Garbino Trash Can

Karim Rashid for Umbra, 1997

❶ The Garbino, a play on *garbage* and legendary silent film actress Greta Garbo, manages to combine the typically contradictory notions of "garbage" and "glamour" into a best-selling product deemed worthy of "art" status by numerous museums. As with most of Rashid's work, the goal was to elevate the mundane, and to do so at a price point affordable to the masses. Karim Rashid comments: "I did at least fifty renderings of receptacles. I felt that banal objects need life, they need presence, they need to make unpleasant tasks more pleasant. I immediately thought about a more sensual object than we've had before, one that is seductively round, wider at the top than at the bottom so that it seems to undulate. Something with a wide, beautiful, inviting mouth to gracefully accept waste."

❷ The Garbino is the "Mini-Me" of the much-heralded Garbo wastepaper bin, reduced in size to appeal to Japanese and European consumers who found the Garbo too big. The form is curvy and feminine, looking more like a female corset than a garbage receptacle—it is the Coca-Cola contour bottle of trash cans. The form is simple but geometrically rich, designed to be a trash receptacle that people want to display rather than hide in a kitchen cabinet. At a slight perspective, the cutouts look like eyes, akin to the cutouts of a Mardi Gras mask, adding to the personality and mystique of the form. Originally available in a variety of translucent colors, it appears that some of Rashid's "sensuous minimalism" has been traded for a slightly less sensual, less translucent degradable plastic, a decision likely influenced by the harsh and often acidic criticisms directed at Umbra and Rashid by sustainability advocates.

❸ Elliptical handles are cut from the two crests created by the saddle-shaped top. The slopes of the ellipses neatly follow the slopes of the curved crests, leaving an even, 1" margin along their lengths, which also happens to be the vertical diameter of the ellipses. The bottoms of the ellipses align with the low points of the receptacle's lip, while their tops reside well above, preventing hands from coming into contact with refuse. The careful attention applied to aligning these key elements simplifies and purifies the form, while also enabling a multitude of complex but harmonious geometric references: parabolic body, saddle-shaped top, elliptical handles, and concave bottom.

❹ The body is smooth and featureless, with the exception of a small seam joining the rounded bottom. The rounded bottom is a thoughtful addition to the design in terms of facilitating cleaning, as rectangular containers invariably accumulate grime and bacteria in the corners. It does raise the question, however, of why designers continue to operate under the assumption that people dutifully clean their trash cans—or fill their trash cans only with materials that leave no residuals. Generally, users live with grimy trash cans, occasionally hosing them out until they are too disgusting to be salvaged, at which point they are discarded, or they diligently line their containers with trash bags. In the former case, a translucent material rapidly loses its appeal for obvious reasons. In the latter case, the aesthetic is all but lost when the form is draped with a trash bag.

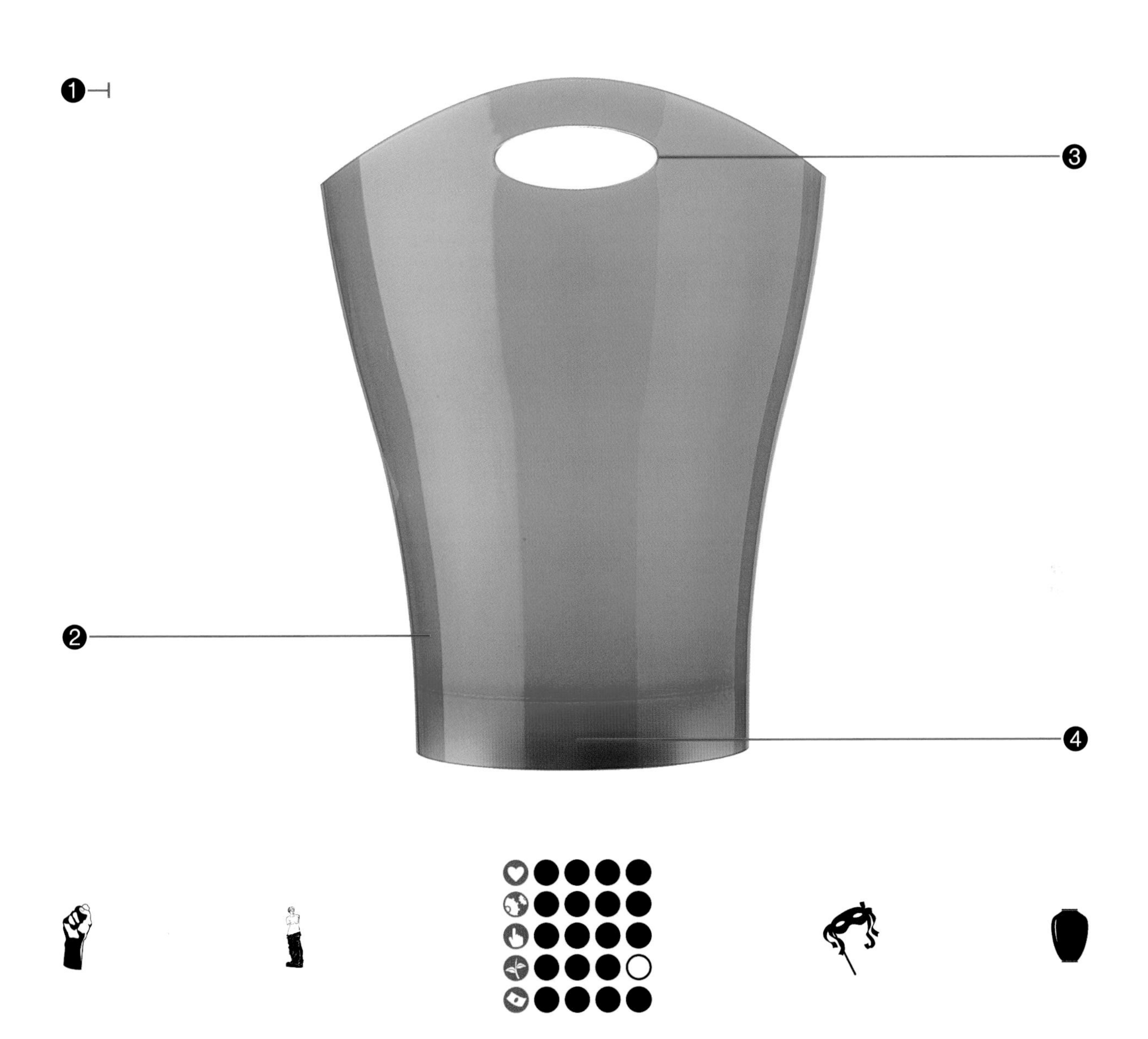

Glock 17 Pistol

Glock, 1982

❶ Considered one of the great firearm innovations of the twentieth century, the Glock 17, a designation derived from the gun being Glock's seventeenth patent, was developed in a competition to supply the Austrian military with a new duty pistol. Prior to this competition, Glock's company had no experience with handgun design, but with manufacturing plastic kitchen boxes, ammo belts, shovels, and utility knives. What possessed Gaston Glock, an engineer by training, to believe that he could create a world-class, combat-quality handgun? "That I knew nothing was my advantage," he says. Glock began work on his gun in 1980, purportedly test-firing prototypes with his left hand so that if one blew up, he could continue drawing with his right. The Glock 17 won the Austrian military competition and received the army's order for twenty-five thousand guns in 1983. Today, Glock is one of the leading manufacturers of handguns worldwide.

❷ The form is simple, utilitarian, comprised of only thirty-four parts—the basis for its legendary reliability. Its envelope is compact and lightweight, the latter attribute owing to its unique polymer frame. The frame resists impact and distortion more effectively than steel or aluminum, resists corrosion and light reflection, effectively absorbs recoiling forces during firing, and is less expensive to manufacture than its metallic counterparts. The Glock 17 is not the first handgun to employ a plastic frame, but it is the first handgun to employ a plastic frame successfully. The pistol does introduce a number of other innovations. For example, the Glock 17 is striker-fired, which means that it does not have a hammer. An internal striker is pulled and released by the trigger action alone, blurring the line between traditional definitions of single-action and double-action.

❸ The pistol's most distinctive feature is its "Safe Action" trigger system, which is employed in lieu of an external safety. The system consists of two triggers, a thin inner trigger contained within a wide outer trigger. The "Safe Action" refers to the fact that the outer trigger cannot be engaged unless the inner one is fully depressed, and therefore the gun cannot be fired unless both are pulled from their centers. This two-step mechanism enables users to carry the gun in a "fire-ready" mode that is also a "safe" mode, which is not possible with traditional designs. Two additional internal safeties are deactivated at different stages along the trigger pull and automatically reactivated when the trigger is released. The total system, then, involves three independent safeties that activate and deactivate in cascade fashion each time the gun is fired. Despite this, unintentional discharges do occur, though investigations invariably reveal that it is because the user pulls the trigger unintentionally (e.g., pulling the gun out of a holster with a finger on the trigger) versus a failure of the mechanism itself. Glock increased the trigger-pull weight to reduce this class of error, but equally critical is training users about the practical differences in how the Glock is to be used versus traditional handguns. Foremost, repeat three times: "Keep your finger off the trigger until you are ready to shoot."

❹ The grips are an extension of the frame, thus additional screw-in grips are not required. The reduced width makes room for a wider magazine—meaning more ammunition—without adversely impacting the circumference of the handle. Grips are contoured and textured to ensure that the gun can be securely held even in adverse conditions, and their symmetric forms support ambidextrous use.

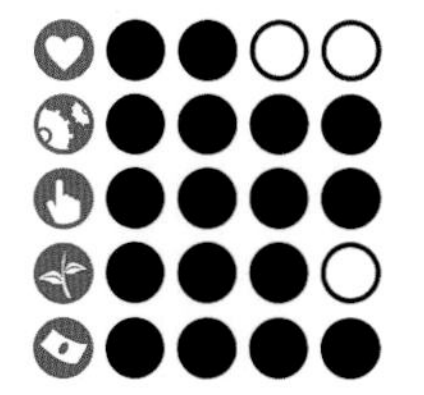

Good Grips Peeler

Smart Design for OXO, 1990

1. The Good Grips Peeler was the first in a series of now widely heralded kitchen utensils that applied universal design principles to create products usable by everyone. Inspiration for the product occurred when Sam Farber, founder of OXO, observed his wife having difficulty using a peeler, due to arthritis. This prompted an extended conversation with David Stowell, president of Smart Design, about the inadequacies of commonly available kitchen utensils. The resulting product was not only received well both commercially and critically, it has made OXO the standard bearer for universal design. Davin Stowell comments: “No one wants to be singled out as having special needs. The peeler was designed to have universal appeal. Those without special needs get an added bonus … We use design to communicate, to give people a better understanding. Sam Farber, OXO’s founder, always insisted that it was not enough to just make something work better. The customer needs to understand it right away.”

2. The peeler looks like a high-quality precision implement, betraying its relatively inexpensive price point. The form is replete with geometric relationships. The handle is divided into thirds. The top third is comprised of vane cuts meant to cushion the key finger pressure points, while the hole in the handle is centered within the lower third. The distance from the top of the grip to the center of the grip hole is the same as the length of the arched cap. These relationships give the peeler balance and proportional coherence. The affordance of the rubber handle is without parallel, and is comfortably held by men, women, children, and the elderly with either the left or right hand. The swivel head blade makes quick work of most any fruit or vegetable.

3. The peeler blade is double-edged and swivels with the pull or push of the stroke, much like the pivoting heads of modern razors that enable users to maintain a proper cutting angle across curved surfaces. The pitch of the blade is precisely set to yield a cut between 0.8 to 1 mm, so only the skin of the fruit or vegetable is removed. The blade is suspended across a glossy black plastic arch support, which contrasts nicely with the matte finish of the rubber. The arch has obvious aesthetic benefits, but less obvious is how it can be flexed open to easily remove and replace the blade. This operation is simplified by depression at the base of the tip, which allows the thumb to make the flex without fear of slippage. It also makes for an effective scraper of dangling peels.

4. The key innovation lies with the grip, which has come to define OXO products. The grip is large and wide, enabling weaker hands to maintain a good grip. The entire handle is composed of soft rubber. It looks grippy, feels grippy, and stays grippy even when wet. Vanes cut into the sides of the neck lead the index finger and thumb to their proper peeling positions, and then lie down to soften the grip when pressure is applied. The OXO logo is blind embossed on the underside of the grip. Its subtlety reinforces the perception that this is a well-designed product—the manufacturer lets the product speak for itself, and chooses not to disrespect the design with brand graffiti. A hole in the handle implies that this is a tool, a product to be displayed and used, versus hidden away in a drawer.

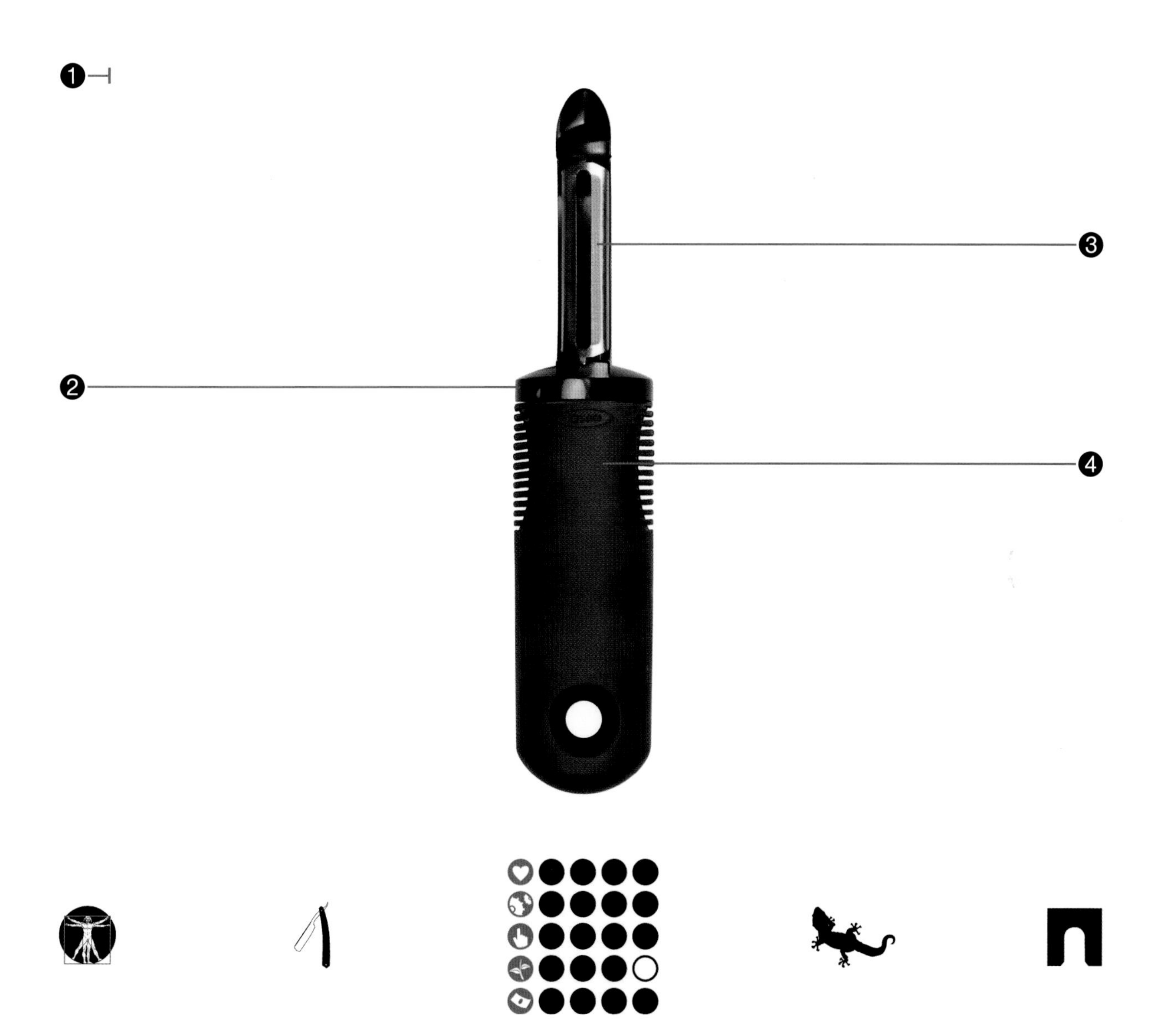
1
2
3
4

HomeHero Kitchen Fire Extinguisher

Peter Arnell for Home Depot, 2007

1. About one third of home fires start in the kitchen. Unfortunately, few kitchens have fire extinguishers—and when they do, they tend to be stored out of sight due to their general unsightliness. While a hidden fire extinguisher may prevent aesthetic clashes with stainless-steel appliances and granite countertops, it is a poor choice in terms of emergency response. In panic situations, people get tunnel vision, and a fire extinguisher that is out of sight is a fire extinguisher that is likely to be out of mind. One solution: redesign the fire extinguisher to be a beautiful object that is proudly featured in the kitchen, rather than hidden away. Peter Arnell comments: "We needed to make something that people would be proud to have out on their countertops ... We tried to create the anti-emergency language with this. The calm aesthetic has been designed to create a balance between the emergency that's happening and some kind of balancing leverage in your mind."

2. The form is small, clean, and simple. Its small size and large cutout handle enable the extinguisher to be used easily with one hand. The characteristic red color has been traded for glossy white, which seems to evoke objections among those concerned with safety, though there is no official standard in the United States for the color of fire extinguishers. The gripping surfaces around the trigger and handle are light gray, à la the iPod, which was an inspiration for the design. A rubber foot of like color has been added to the bottom to protect those fragile granite countertops. Challenging institutions and conventions that have not fundamentally changed in over one hundred years is no small task, but the reward is the most beautiful fire extinguisher ever created.

3. The red safety pin locks the trigger until pulled. Its red color draws the appropriate amount of attention, saying effectively, "Start here." The circular ring clearly affords pulling. This prerequisite action inclines users to grab the extinguisher with the dominant hand at the correct location from the correct orientation, ensuring that the hand is immediately ready to depress the trigger and extinguish the flame.

4. The handle is a cutout in the form, lined with an ergonomic gripping surface. The product name is blind debossed just below the safety pin, a subtle and elegant marking. The gripping affordance is clear, and the trigger is positioned for thumb activation. The device is held easily and feels well balanced. As there is no nozzle, the device can be used with one hand, leaving the other hand free to address other critical issues, like turning off gas, flipping a breaker, or calling emergency services.

5. The white color choice seems a more logical color for a fire extinguisher than red—that is, white relates to coolness and emptiness, whereas as red in this context relates to heat and fire. However, nuanced logic like this is generally too complex to process in emergency situations, and it is certainly no match for the RED = FIRE and RED = EMERGENCY associations that are strongly conditioned in Western minds. This, coupled with its thermos-like form factor, has surely raised the hackles of many a fire and safety specialist. Deviating from the iconic aspects of traditional fire extinguishers risks making the product less recognizable as a fire extinguisher, and potentially less viable as a safety device. However, if users are reluctant to buy and accessibly locate traditional fire extinguishers in their kitchens, the trade seems inarguable.

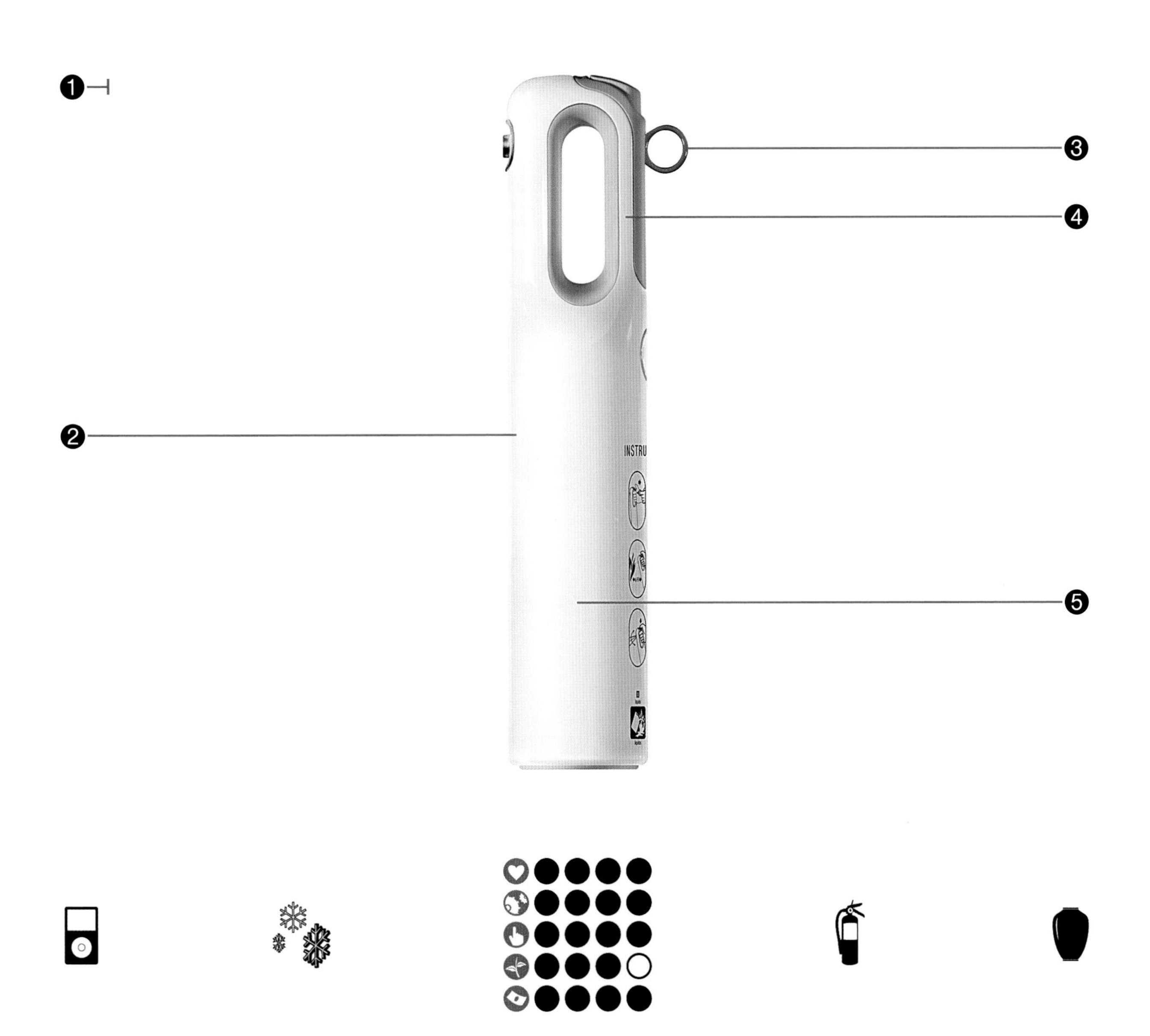

Hug Salt and Pepper Shakers

Alberto Mantilla, 2002

❶ Perhaps a response to the racially divisive midcentury Aunt Jemima and Uncle Mose salt and pepper shakers, or more symbolically a statement about how superficial differences can be overcome through gestures of kindness, the Hug Salt and Pepper shakers clearly have more to do with delivering a progressive social statement than delivering kitchen table condiments. Alberto Mantilla comments: "The very nature of the Hug connotes brotherhood. The bold use of black and white suggests that we are all brothers and sisters on this planet and we need to treat each other with kindness, compassion, and respect."

❷ The forms are abstractly anthropomorphic, resembling two ghosts in an embrace. The shakers are identical except for their colors—black for pepper, white for salt—with three dispensing holes in the eyes and mouth position. While some may find salt and pepper flowing from the eyes and mouths of these two ghoul-like figures to be a bit too horror-genre for their tastes, the oddity is quickly overcome as the majority of their time is spent standing and hugging versus shaking and spewing condiment. The affective response to the shakers is interesting. One finds the need to keep them together, in the "hug" position, as if not wanting to leave the unused shaker "hanging." It is hard to imagine evoking this kind of reaction with nonanthropomorphic forms. Everything about their shapes is roundy, drawing on innate baby-face cognitive biases. These biases confer attributes like cuteness, naïveté, and helplessness to the shakers, attributes that will be more strongly felt by women—which is appropriate, as women are the target demographic for this product. Functionalists will point out that the "hug" position minimizes the amount of table real estate required, but this is incidental to the design.

❸ One of the most compelling aspects of the product design it its *proposition density*—that is, the number of propositions or messages expressed by the form factor. For example, if a product clearly expresses how it is to be used, this is one proposition. If it expresses affective information, that is another proposition. And so on. The more propositions expressed by a product form, the greater its propositional density. When a product has high propositional density, and the propositions align with one another, a synergistic aesthetic is achieved. In the case of the Hug shakers, the propositional density is very high and very synergistic: colors convey contents, anthropomorphic forms convey persons, holes convey condiment exits, holes convey facial features, anthropomorphic forms convey ghosts, symmetric forms conveys equality, roundiness conveys baby-like attributes, hugging conveys harmony, anthropomorphic forms and colors convey race, anthropomorphic forms and colors and hugging convey racial harmony, colors and yin/yang (as viewed from above) convey balance and harmony, and there are no doubt many more. With the exception of the ghost and ghoul associations, all of the propositions expressed by the salt and pepper shakers are coherent and consistent with one another, creating a meta-aesthetic that is greater than the sum of its parts. This kind of aligned propositional density endows a product with a depth of expression that reveals itself slowly over time—more like a story than a product—sustaining interest in its aesthetic and its expression of social commentary. It is rare for even complex products to express this many propositions. That so much is expressed by so little while remaining so simple and subtle is a remarkable demonstration of communicative design.

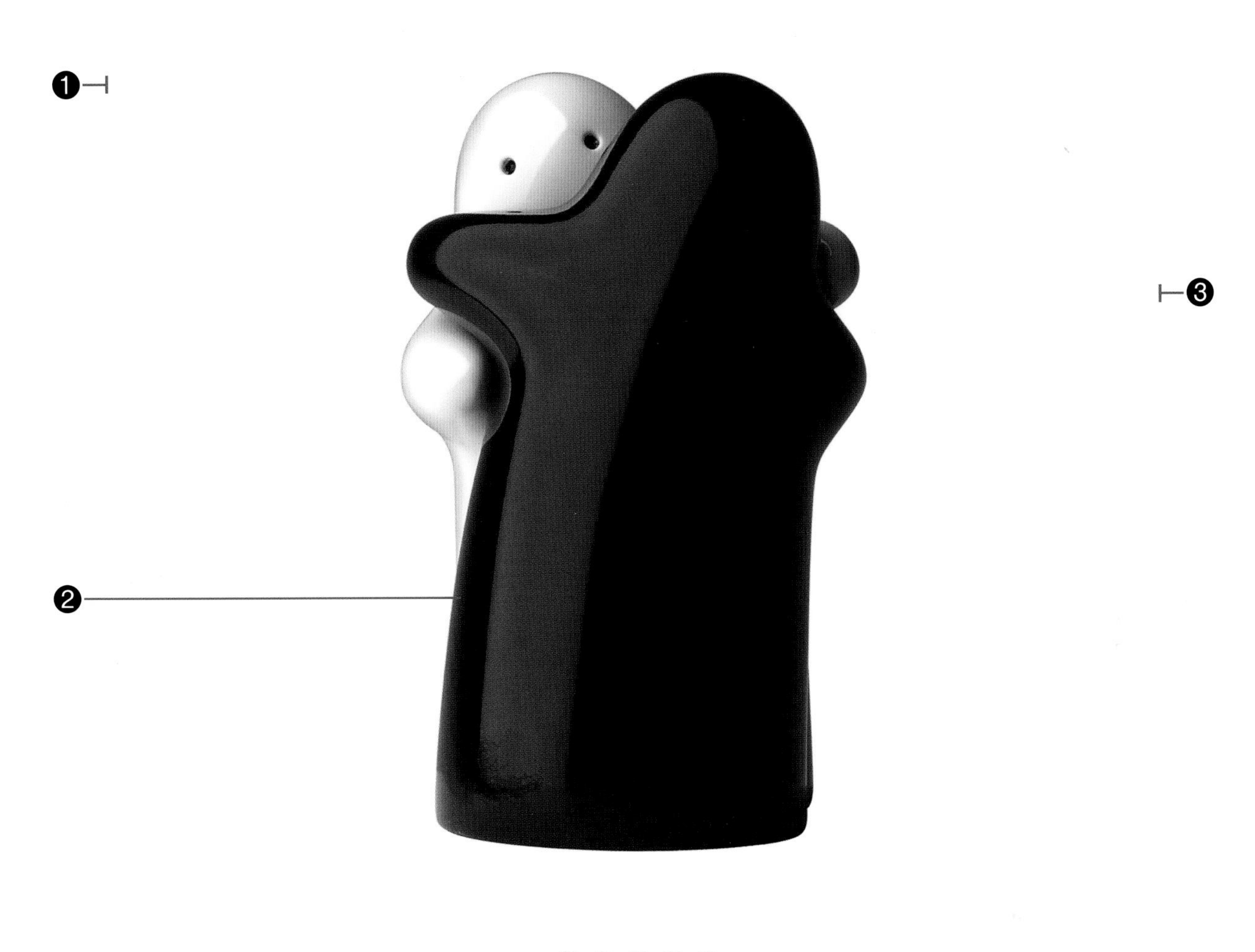
1
2
3

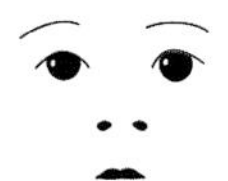

HurriQuake Nail

Ed Sutt and The Stanley Works, 2005

❶ For want of a nail, a kingdom was lost—for want of a HurriQuake nail, property damage from hurricanes and earthquakes is roughly twice what it would be otherwise, a cost totaling tens and perhaps hundreds of billions of dollars annually. How is it that a mere nail could make this kind of difference? The basis of virtually all modern construction is joining wood with nails to form structure. If the join, wood, or nail fails, so too does the structure. The key is identifying the weak link. Ed Sutt comments, "I had the opportunity to look at hurricane damage firsthand in 1995 when Hurricane Marilyn hit the U.S. Virgin Islands. I saw the impact, the destruction caused by the storm, and the difficulty that people had in rebuilding because everything had to be brought by boat. I also realized that the fasteners or nails were the weak link in the chain."

❷ A problem created by the ring shanks on the lower part of the nail is an enlarged hole, which makes the upper nail fit loosely. The problem is remedied by increasing the diameter of the shear shank, which not only fills the void and tightens the joint, but also reinforces the nail against side-to-side shearing, a common stressor in earthquakes. Additional shear resistance comes from metallurgical advances in the carbon steel alloy, which balances rigidity and pliability. For all of its innovations, the nail is not that visually distinctive to the untrained eye—perhaps a colorized zinc coating (e.g., red) should have been considered to give the nail a more recognizable and distinctive identity.

❸ The head is 25 percent larger than traditional nails, reducing pull-through, a common form of failure that occurs when a join is pulled apart and the nail head is pulled through the wood. The tradeoff is compatibility with common nail guns, meaning that the designers had to balance the goal of attaining maximum head surface area with that of supporting nail gun standards. The enlarged head also allows simple labeling, HQ1–HQ4 for varying lengths and diameters, facilitating code inspections.

❹ Just below the head, a fluted spiral shank slows the rate of nail entry at the final point of insertion, reducing overdrive of the nail into the wood and locking it into place with a twist.

❺ The lower part of the nail is covered in ring shanks, its most visually distinguishing feature. Ring shanks are essentially one-way barbs, easy to put in and hard to pull out. The ring shanks address another common form of failure known as pullout, which occurs when the join fails and the nail is pulled out as the wood separates. A common user complaint, in fact, is the difficulty of removing the nail in the event of error or demolition. Once joined, removing the nail is nontrivial.

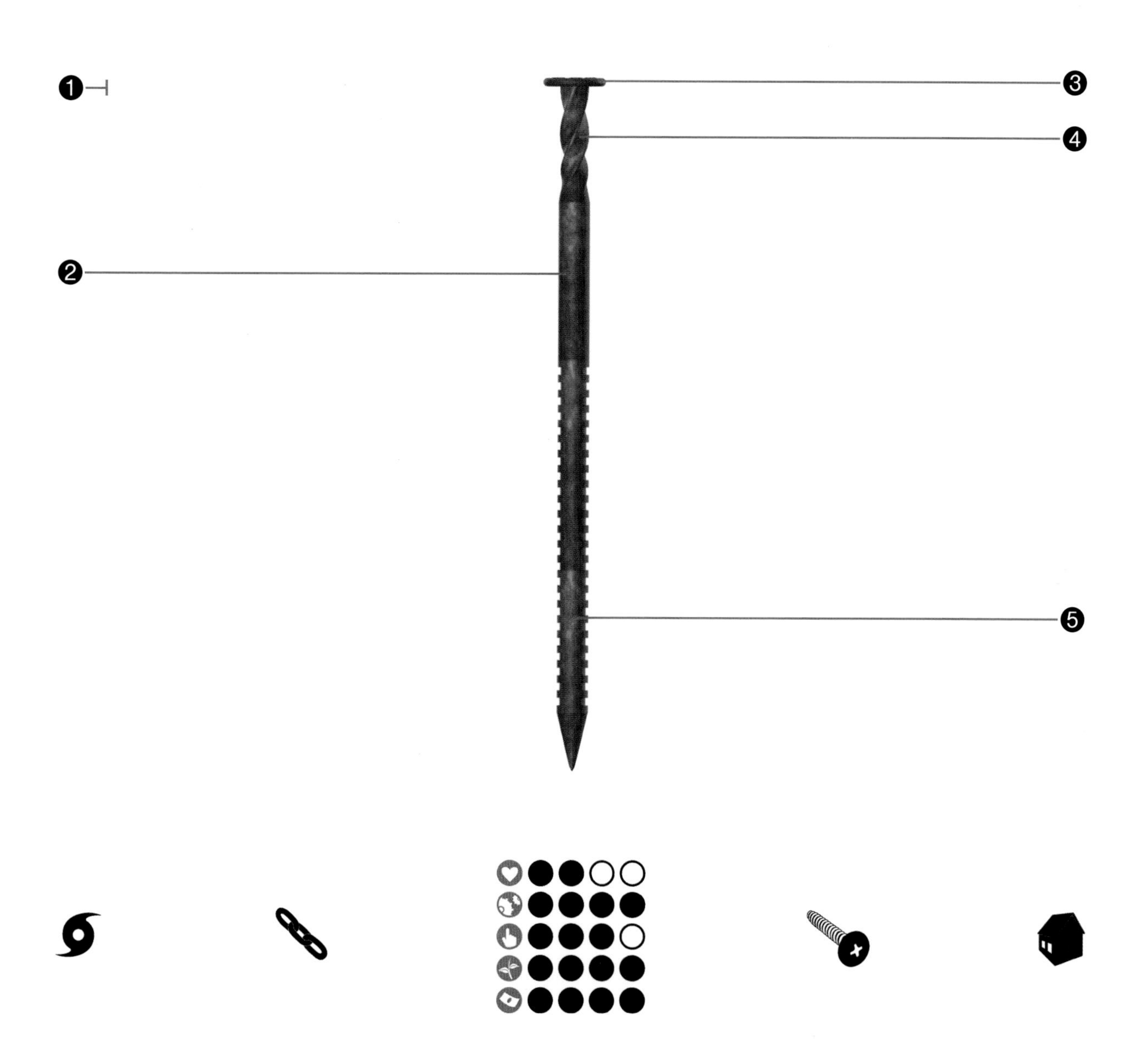

1
2
3
4
5

iBOT Mobility System

DEKA Research and Development, 2001

❶ On rare occasion, a product comes along that addresses deep *user needs*—needs that users themselves would be unable to articulate if asked—and in so doing radically redefines the norms and expectations for a product category. The iBOT, ancestor to the Segway, is one such product. With its ability to raise occupants to standing eye level, climb stairs, and traverse rough terrain, the iBOT redefined what a wheelchair can and should be. How did iBOT designers attain such a deep understanding of the needs of the wheelchair bound? Dean Kamen comments: "I try to understand the basic laws of nature. Beyond this, I do very little research as to what the product should be. You would never get the iBOT by doing research on wheelchairs. If you do 'product research,' the product that you end up with will be similar to what already exists... You have to start with basic questions: if this person is now missing this amount of functionality, is there some alternative to a wheelchair that is both dramatically better and not prohibited by the laws of physics and the current state of engineering and technology?"

❷ In standard mode, the iBOT looks like a conventional motorized wheelchair. It is in balance mode that the product literally blossoms, vertically folding and stacking its wheels to lift the occupant to standing height. The iBOT in this configuration is a true spectacle, tall and precarious looking, but amazingly stable (even when pushed), a testament to the coordinated actions of its numerous sensors, gyros, and motors. User testimonials speak to the psychological impact of being able to interact with others at standing eye level using the iBOT, making it clear that the product elevates more than just bodies.

❸ Two seating options are available: rehab seat and automotive seat. Both seats are well padded, and considered by users to be on par with other motorized wheelchairs. The seat back, arm rests, lap belt, calf panels, and foot pedals are all adjustable. A user control panel is located on the right or left armrest as required by the user, and attached to a swing arm to allow closer access to desks or tables.

❹ The iBot has six wheels, four primary and two satellite, which support four modes of operation: standard, four-wheel drive, stair-climbing, and balance. In standard mode, the iBot is faster than most motorized wheelchairs. In four-wheel drive mode, all four wheels engage to traverse curbs and navigate rocky terrain. In stair-climbing mode, the iBot climbs stairs, supporting the user by power-assisting the rotation of its two sets of primary wheels around each other. Although innovative, it is the least usable and least safe of the modes of operation. In balance mode, the iBot lifts users to standing height. The tires are durable, but susceptible to flats. One of the costs of concentrating so much technology into one product is bulk. The iBOT is large, heavy, even for a motorized wheelchair. It fares poorly in small but essential home spaces like bathrooms, and transport essentially requires a truck or van. Another of the costs is a failure to fit into a bureaucratic healthcare system that has checkboxes for "wheelchairs" but not for "robotic mobility systems." Accordingly, the product was classified a luxury item, rendering it generally ineligible for insurance coverage in the United States. With its high price point, $25,000, and lack of insurance coverage, the iBot was discontinued January 1, 2009.

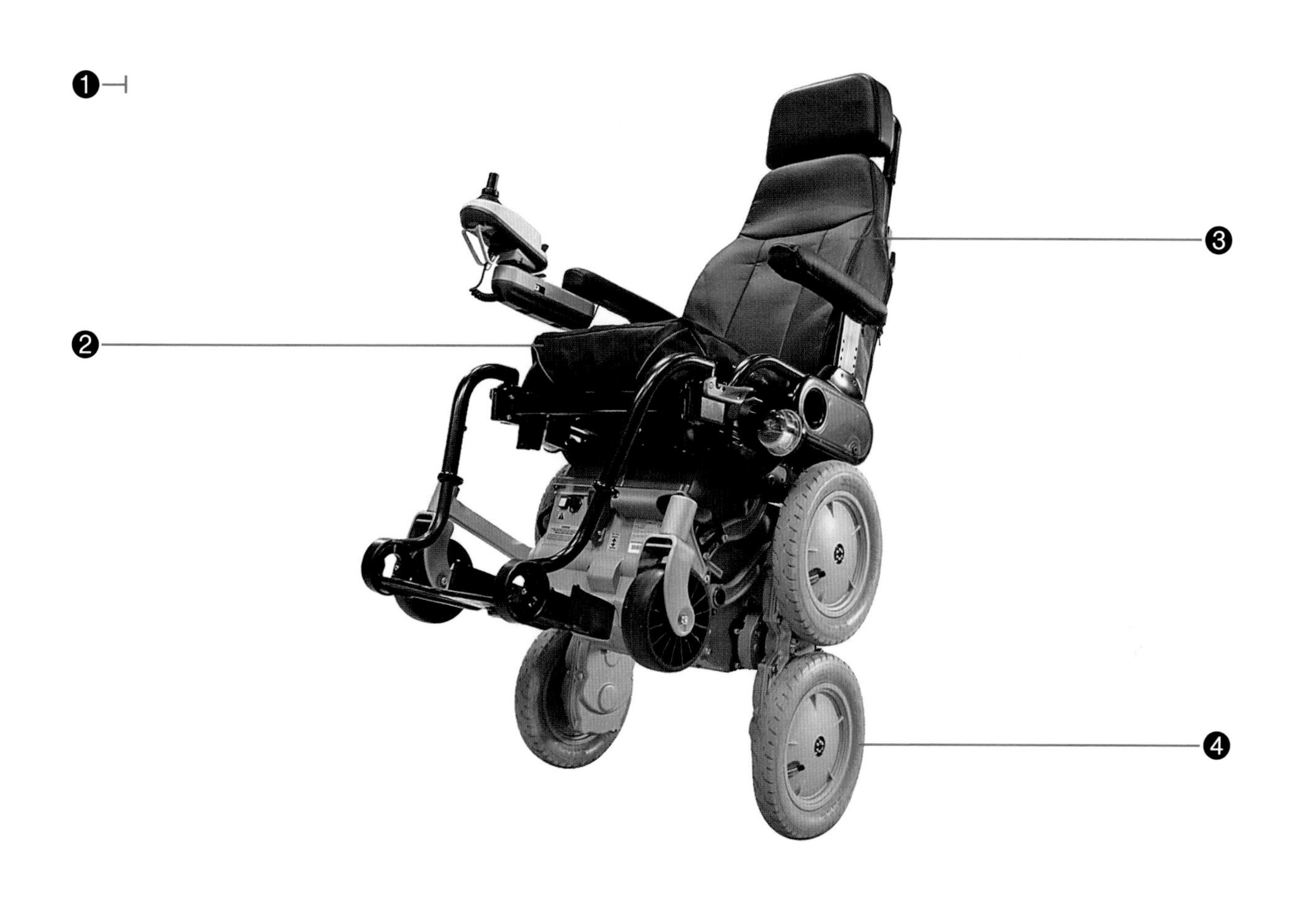
1
2
3
4

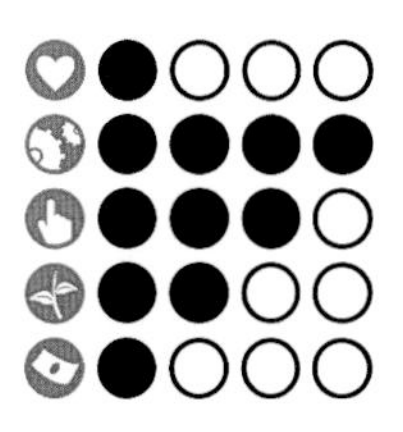

il Conico Kettle

Aldo for Alessi, 1986

❶ Like Aldo Rossi's buildings, the il Conico emphasizes basic geometries in interesting ways, employing a kind of Euclidian recursion to add depth and interest to its primal form. How does a noted architect find inspiration for the design of everyday kitchen objects? Aldo Rossi comments: "I have always had a strong interest in objects, instruments, apparatus, tools ... without intending to I used to linger for hours in the large kitchen at S., on Lake Como, drawing the coffeepots, the pans, the bottles. I particularly loved the strange shapes of the coffeepots enameled blue, green, red; they were miniatures of the fantastic architectures that I would encounter later."

❷ The lid completes the conical form of the body and sits smartly atop the base. At its apex and horizontally centered within the form is a small orb, giving the lid a hatlike appearance and representing the only grippable affordance other than the handle. Unfortunately, like the handle, the heat-conducting orb makes removing the lid off of a kettle of boiling water a perilous activity.

❸ The spout is triangular, and though disproportionately small for the body, works aesthetically—when combined with the lid, it makes the kettle appear avian, like a cardinal or canary. This "beak" is fundamental in creating the positive affect that this kettle arouses, softening its otherwise angular and inorganic aesthetic. The top of the spout flows into the top of the body, an elegant extension of the horizontal line and a logical position for the exit.

❹ The conical body provides ample volume and maximum base surface area for effective heating. The slope of the cone is large, however, requiring a steep tipping angle to pour, which can result in unwanted rushes of hot liquid.

❺ The handle is a thin bar of stainless steel bent to form an inverted right triangle, the top of which is set to align with the top of the kettle body and spout, and the side to align with the perimeter of the base. The resulting lines, both explicit and implicit, are key in creating a number of secondary geometries. Functionally speaking, that steel is an effective conductor of heat seems to have been lost in process somewhere—welding a stainless-steel handle to a stainless-steel kettle destined to boil water borders on negligence. One wonders if anyone used this product prior to its manufacture.

❻ The conical form is simple and geometrically pure—almost Platonic—forming an equilateral triangle in profile. As Kimberly Elam has noted in her book, *Geometry of Design*, the form abides by the Rule of Thirds and can be effectively analyzed using a 3x3 grid. The top row is comprised of the lid; middle row of the spout, middle base, and handle; and the bottom row of the wide lower base. The left column is comprised of the spout and left base; middle column of the lid and middle base; and right column of the handle and right base. In this framework, secondary geometries are also evident—elements in the bottom-right two-thirds form a square, the spout is half the height of the middle row, and so on. The primary horizontal line—the line that runs from the top of the spout across the top of the body to the top of the handle—cuts at a proportion approximating the Golden Ratio. Material and finish are uniformly stainless steel.

1
2
3
4
5
6

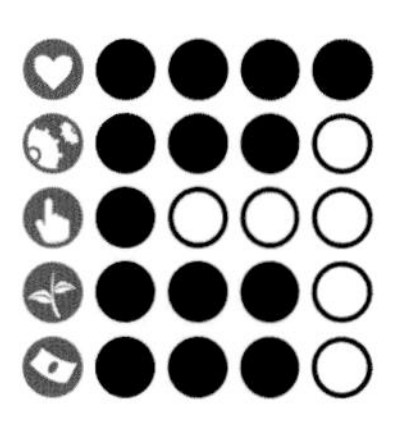

IN-50 Coffee Table

Isamu Noguchi, 1944

❶ It is difficult to believe that a table as simple and elegant as the IN-50 could have in any way been inspired by drama or negativity, but history teaches that great design often finds inspiration from unlikely sources. Isamu Noguchi comments: "I went to Hawaii in 1939 to do an advertisement (with Georgia O'Keefe and Pierre Roy). As a result of this I had met Robsjohn Gibbings, the furniture designer, who had asked me to do a coffee table for him (I had already done a table for Conger Goodyear). I designed a small model in plastic and heard no further before I went west. While interned in Poston I was surprised to see a variation of this published as a Gibbings advertisement. When, on my return I remonstrated, he said anybody could make a three-legged table. In revenge, I made my own variant of my own table, articulated as the Goodyear Table, but reduced to rudiments. It illustrated an article by George Nelson called 'How to Make a Table.' This is the Coffee Table that was later sold in such quantity by the Herman Miller Furniture Company."

❷ For this variant of the Goodyear Table, Noguchi refined the base by reducing complexity and increasing symmetry while still maintaining an organic, nongeometric form—telling changes that suggest the basic principles Noguchi applied to improve the design. Its affect is soothing, meditative, a table you would find in a Japanese garden. The table is comprised of three pieces, but only two basic forms: a triangular glass top and two symmetric base pieces interlocked by a single metal dowel. The base pieces function together as a tripod, though it would never be identified as such.

❸ The table is one of the purest expressions of minimalism and organicism in furniture design. The tension in the form is a common theme in Noguchi's work, and here it alternates between harmonious balance and imminent collapse. The base pieces sit in opposition, creating a kind of visual yin/yang. All edges are rounded, giving the elements the appearance of smoothed river stones. The tabletop's amoeba-like shape and position on the base pieces are slightly off, irregular, very wabi-sabi. The sum effect will make you want to serve sake and prune Bonsais. The base pieces are also romantically anthropomorphic, each looking like an abstract person, one bending over the other, mouths joined, with their privacy betrayed by a transparent cover. The bases are available in ebony-finished poplar and natural-finished walnut, but the ebony is the more popular of the two. This is likely because it is a more natural compliment to the Zen-like aesthetic of the table, the ebony pieces looking like slabs of stone versus cut wood. While functional as a table, it is clear that the IN-50 is sculpture first, furniture second. Accordingly, to understand the design, it is important to understand the sculptural philosophy of the master himself. Noguchi comments: "Sculpture may be anything and will be valued for its intrinsic sculptural qualities. However, it seems to me that the natural mediums of wood and stone, alive before man was, have the greater capacity to comfort us with the reality of our being. They are as familiar as the earth, a matter of sensibility. In our times we think to control nature, only to find that in the end it escapes us. I for one return recurrently to the earth in my search for the meaning of sculpture... Sculpture is the definition of form in space, visible to the mobile spectator as participant. Sculptures move because we move."

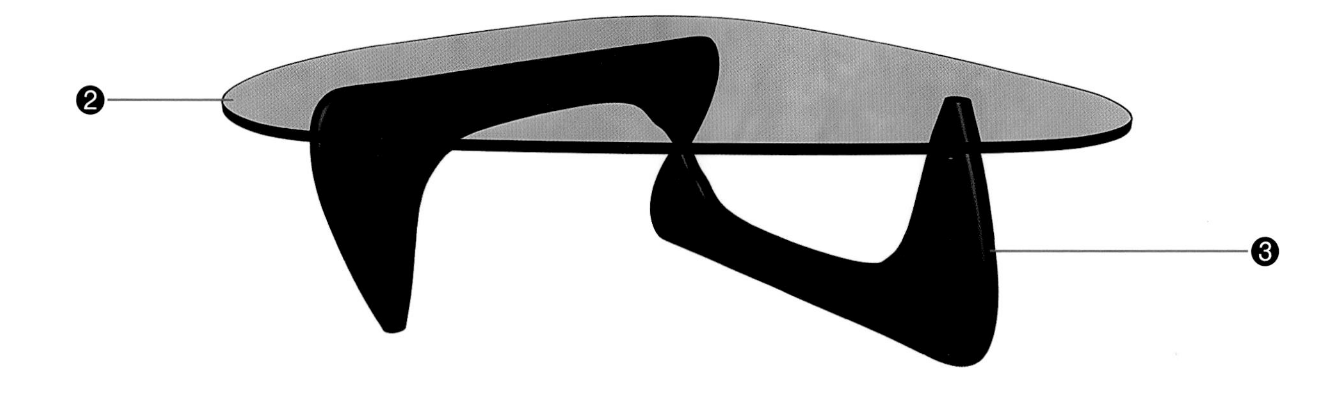

iPhone

Apple, 2007

❶ Apple's foray into mobile phones has set the user experience bar for telephony generally, not just mobile phones. From a feature perspective, the iPhone offers little new relative to its competitors. Its true innovation lies in the way it integrates technology to address deep *user needs*, versus merely accommodating superficial *user wants*, underscoring the difference between the two and affirming Don Norman's observation that "when technology delivers basic needs, user experience dominates." Jonathan Ive comments: "The word *design* is everything and nothing. We think of design as not just the product's appearance, it's what the product is, how it works. The design and the product itself are inseparable."

❷ The control surface employs a capacitive touch screen, which is intuitive and responsive to all but those with long fingernails. Using the display as the primary control surface maximizes content real estate and enables the user interface to be optimized for individual applications. The Multi-Touch interface effectively uses intuitive touch-based gestures for interaction. The initial user experience—sliding a control to unlock the phone—simultaneously introduces gesturing to the user and connects with them on an emotional level, like a handshake.

❸ The icon used for Home is appropriately modern, a square with rounded corners, rather than the more common and banal house icon. Absent any other options, the face affords pressing the button, which is the primary method for waking the phone from its sleep state and gives users an ever-visible fail-safe option to return to the start position in case of trouble.

❹ The form is dominated by the face, which is a seamless pane of glass except for a single button—the Home button. The size of the iPhone is perfect for holding, but less so for users accustomed to carrying their phone in their pockets. The display and primary control surface is centered between two black regions of symmetric proportion, one containing the speaker and the other the Home button, both centered within their regions. The face is framed in metal with a chrome finish. The frame accentuates and protects the black face, but the cost is stylistic clash and garish ornamentation, especially when viewed against the minimalist face and matte aluminum body. Quick access controls are located on the upper perimeter, and include volume, ringer on/off, and power sleep/wake.

❺ Application icons are clear and colorful, benefiting from the high-resolution display and appropriately sized for fingers. Interactions are rich with visual transitions: applications zoom in from a point when selected and screens slide when paged. Lists scroll and screens page in proportion to the rate of user action—the faster users flick their finger, the faster the display moves. Additionally, objects have Newtonian-like momentum, gradually decelerating once a gesture is complete and bouncing back or "rubber-banding" when boundaries are reached. These are more than interesting effects (though they are those); they are antidotes to a problem intrinsic to small, multi-functional displays: disorientation.

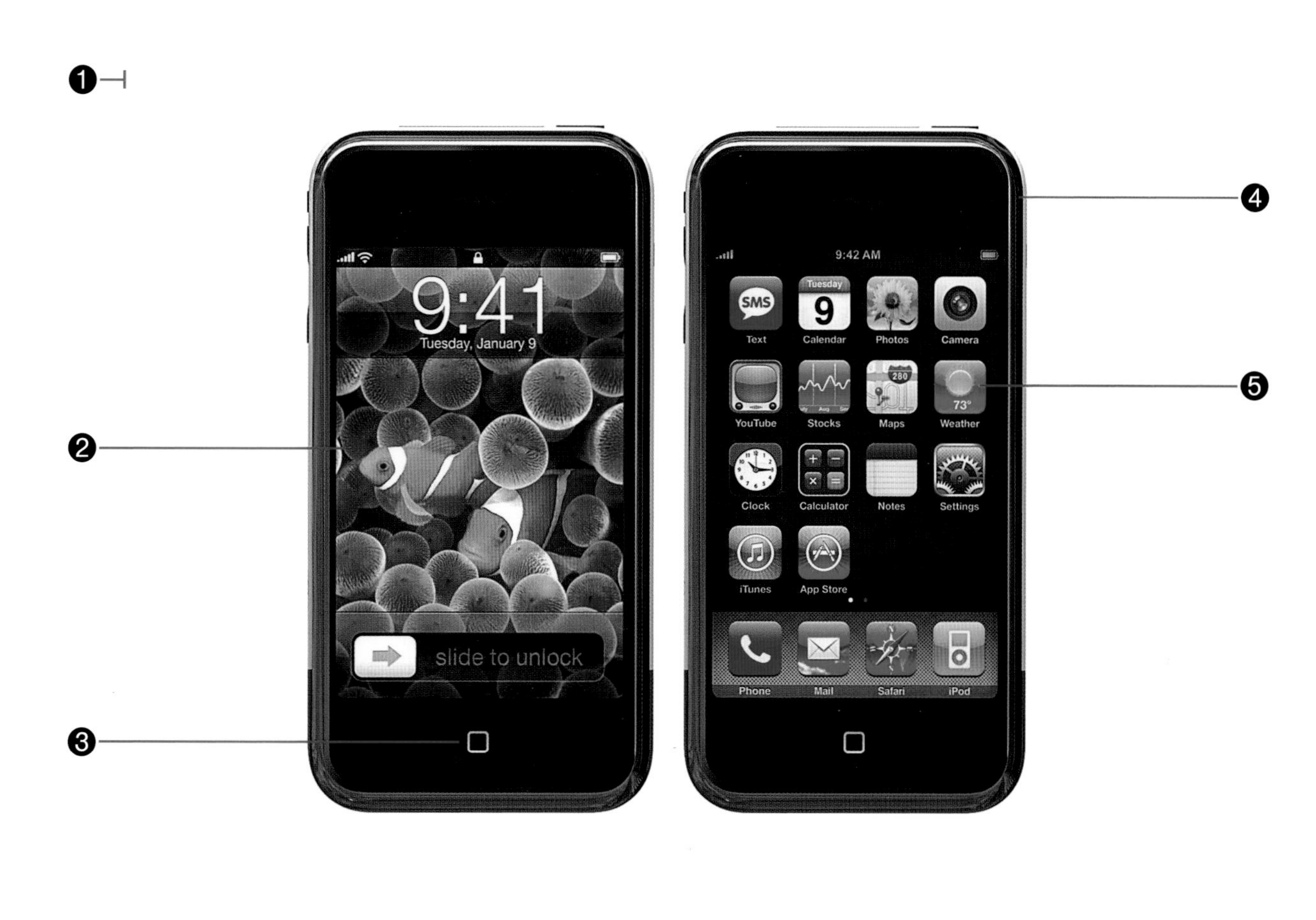

9:41
Tuesday, January 9
slide to unlock
9:42 AM
Text
Calendar
Photos
Camera
YouTube
Stocks
Maps
Weather
Clock
Calculator
Notes
Settings
iTunes
App Store
Phone
Mail
Safari
iPod

iPod 5G

Apple, 2005

❶ The Apple iPod has become the symbol of a generation. Its simplicity of design and flawless integration with the iTunes Music service has forever transformed the music industry while also demonstrating the power of superior design and innovation in business. Jonathan Ive comments: "Steve [Jobs] made some very interesting observations very early on about how this was about navigating content. It was about being very focused and not trying to do too much with the device—which would have been its complication and, therefore, its demise… What's interesting is that out of that simplicity… came a very different product. But difference wasn't the goal. It's actually very easy to create a different thing. What was exciting is starting to realize that its difference was really a consequence of this quest to make it a very simple thing."

❷ The case is seamless and modern, absent any visible screws or fasteners. Its form is composed entirely of rectangles and circles. Given its recursive simplicity and compositional elegance, it is almost unforgivable that the clickwheel is not vertically centered within its space. The deep finish resembles glass or porcelain, and the material is similarly nondurable. The corners are rounded to prevent chipping. Though often criticized for its fragility as a mobile device, this attribute may actually reinforce the perception of the iPod as a precious object—a special item to be cared for and adorned with protective accessories. The voluntary act of caring for an object in this way often promotes an emotional connection with the owner, increasing its perceived value. To supplant this bonding process with a more ruggedized iPod would reduce the caregiving role of the owner and, correspondingly, the perceived value of the object.

❸ iPods are available in white and black only. Objects that use few colors are generally perceived to be of higher value than highly colorful objects. In the extreme, colorless objects—that is, white and black—are accorded the highest values, often using two tones or two textures to achieve contrast among their elements.

❹ The click wheel enables both scrolling and playback control. Scrolling is achieved by lightly rubbing the wheel in a circular motion—a motion, in some instances, resembling a caress, which may help explain the appeal of the interaction. Rotary controls of this type are easy to use and particularly effective at navigating n-length dimensions using minimal control surfaces. Subtle variation in the tone and texture of the wheel implies a use distinct from the case. Playback control is achieved by pressing the wheel at the marked locations. The center button is used for item selection. Both the click wheel and center button provide clear, haptic feedback when pressed. It is interesting that the design uses a label for the menu function and icons for the playback functions, choosing clarity of function over consistency. The elegant integration of these functions presents significant capability with an economy of form.

Now Playing
6 of 15
Feel Good, Inc.
Gorillaz
Demon Days
3:41
-1:20
MENU

Jaipur Foot

Ram Chandra, P.K. Sethi, and Devendra Raj Mehta, 1969

❶ Great product innovations are often the result of professional designers employing methodical research and design practices to solve problems, but they are as often the result of improbable people employing unconventional practices in unlikely circumstances. The latter is the case with the Jaipur foot. Ram Chandra, an Indian sculptor hired to teach art to polio victims at a local hospital, observed the many problems and correspondingly high rejection rates of Western prostheses by amputees. While thinking about alternative designs that addressed Eastern needs, he got a flat tire while riding his bicycle. Chandra took the tire to a local repair stall, which re-treaded the tire with vulcanized rubber. This seeded the idea of a prosthetic design that used vulcanized rubber, a viable local material that could be supported by local artisans. Chandra consulted with Dr. P.K. Sethi, the physician who had hired him to teach art, to refine the design and partner in its implementation. P.K. Sethi comments: "Wearing shoes, which were integral to the Western designed limbs, was uncomfortable in our hot climate. Our people walk barefoot or in well-ventilated footwear… We are essentially a floor-sitting people, requiring a range of mobility in our feet and knees, which is not needed in the chair-sitting culture of the West. We should not expect our people to change their lifestyle because of a design we were forcing on them." In the West, the average cost of a prosthetic foot is $8,000. In India, the Jaipur foot costs about $30 and is given away free. An extraordinary cost difference with no significant difference in performance—an embarrassment for the Western world, and a reminder that necessity plus creativity trumps resources in fueling innovation.

❷ The foot is made of three blocks: forefoot and heel blocks are made of sponge rubber, and the ankle is made of a block of wood. These blocks are joined and bound in a rubber shell that is vulcanized in a mold to give the appearance of a natural foot. Once fitted, the foot is essentially maintenance free for the duration of its five-year life cycle. The range of performance achieved by exploiting the variable tensile and compressive properties of the simple materials used is impressive, a lesson about what can be achieved when designers can't simply throw technology at a problem. Wearers are able to run, sit cross-legged, squat, climb trees, work in wet and muddy fields, and wear shoes.

❸ Where ends product design? With the object? With the supporting services? Marketing? The question is germane with regards to the Jaipur foot, as the experience of obtaining the prosthesis is as important to its success as the materials from which it is made. The registration and fitting process, designed principally by Devendra Raj Mehta, makes sure that patients are accepted immediately upon arrival, 24/7. The patients and their families, who are characteristically penniless, are given food and drink and quarters to stay in while the fabrication and fitting process occurs. Meals are communal, allowing patients and their families to socialize and commiserate. Once the prosthesis is complete, a process that typically takes one full day, the patients are given fare for the return trip home, and walk upright out of the hospital, with dignity. The process is akin to a rebirth—from lame to fully functional in one day—an experience designed not just to restore patients' mobility, but their psyche as well.

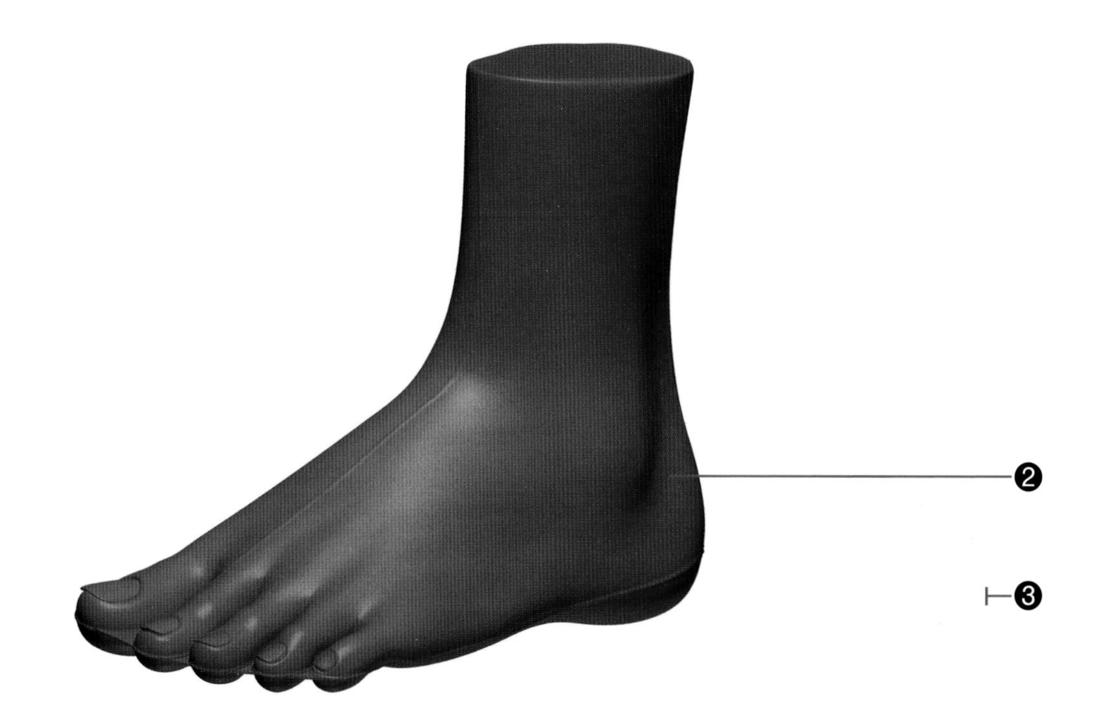

2
3

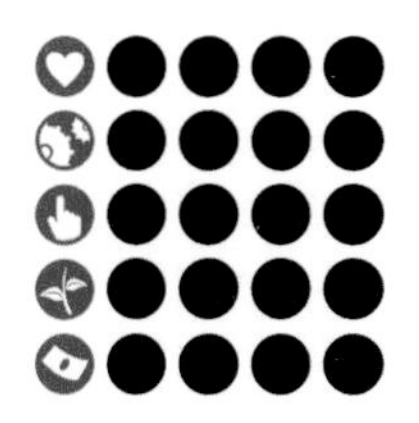

Jawbone Wireless Headset

Fuseproject for Aliph, 2006

❶ As it becomes increasingly clear that many people will, given the option and in spite of the affront to those around them, choose to be connected to friends and family by mobile phone at all times—working, driving, shopping, urinating—so has the need grown for a hands-free wireless communication device compatible with such activities. The Jawbone is perhaps the first mobile phone headset to acknowledge the inevitable transition from occasionally worn mobile phone accessory to permanently worn fashion accessory, and all the stylish implications that this entails. Yves Béhar, founder of Fuseproject, comments: "We really decided to take out all of the techie stuff, all of the nerdy stuff, and try to make it as beautiful as we can. Think about it, the care we take in selecting sunglasses or jewelry or accessories is really important—if it isn't beautiful, it does not belong on your face."

❷ The ear loop is skeletal, itself shaped like an ear. This organic aesthetic conflicts with the machine-like aesthetic of the earpiece, and disrupts the integrity of the design. The earpiece comes with four earbuds and four ear loops, an attempt to accommodate variability in ear size and shape with multiple configurations. This Aeron-like approach to managing anthropometric variability seems sensible enough, but the implementation is often unsatisfying—a lot of twiddling and adjustment that never seems to achieve the right fit. Although offering multiple options like this is increasingly popular with manufacturers, it rarely works in terms of user experience. Better to go with a standard form factor, accommodate variability with pliable materials, and avoid creating the perception of a problem altogether.

❸ The form is modern and architectural, looking like a refined version of Uhura's earpiece from the original *Star Trek*. The Jawbone's styling is more techno-chic than fine jewelry, but it approaches the latter more than any other earpiece to date. The grated exterior surface looks like industrial venting, an interesting aesthetic but one that implies that the device is sufficiently high-powered to need cooling in this way (which it does not). Given the debated association between radiofrequency energy from cellular phones and brain cancer, one would think that a bluetooth device that is to be worn near the brain for extended periods would be designed to downplay, not exaggerate, its perceived power requirements.

❹ Buttons are concealed beneath the grated cover, the pressing of which is the means to their activation. The concealment of all buttons makes for a clean aesthetic, but the cost is compromised usability. For example, when the position of the earpiece needs adjustment, returning it to the correct position can accidentally activate one of the buttons, which, among other things, can result in disconnected calls. Another usability issue regards the basics of putting the earpiece on and taking it off. It is difficult—too difficult. Until users get the hang of it, the time spent fumbling about results in missed calls that go to voice mail.

1

2
Jawbone
3
4

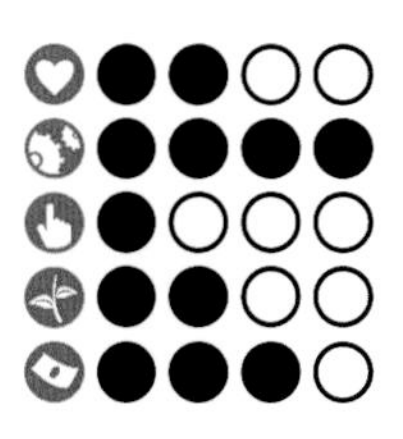

Jelly Fish Watch GK 100

Swatch, 1983

❶ Swatch—an abbreviation for "second watch," to promote the notion that consumers would want more than one, but today generally interpreted to mean "Swiss watch"—played a central role in fueling one of the most dramatic industrial comebacks in history, as well as launched a revolution in product design. Driven to the brink of insolvency by low-cost Japanese digital watches, the Swiss watch industry with its high-precision analog watches seemed doomed to go the way of the buggy whip. Enter Nicolas Hayek. Hayek and his colleagues consolidated and reorganized the company. They retooled processes, simplified the watches for manufacturing efficiency, and emphasized the emotional design of their timepieces to set them apart from the banal digital watches commoditizing the market. The strategy worked, and has restored the company and Switzerland as a major player in the wristwatch market. Nicolas Hayek, cofounder and chairman of the board of directors of the Swatch Group, comments: "I understood that we were not just selling a consumer product, or even a branded product. We were selling an emotional product. You wear a watch on your wrist, right against your skin. You have it there for twelve hours a day, maybe twenty-four hours a day. It can be an important part of your self-image. It doesn't have to be a commodity. I knew that if we could add genuine emotion to the product, and attack the low end with a strong message, we could succeed."

❷ The watch reveals its mechanicals, giving it a dynamism lacking with most analog watches. The technique makes the product feel honest, as if the manufacturer is eager to display their work to consumers for inspection.

❸ Without the translucent strap, the name "Jelly Fish" would simply not make sense. The strap completes the effect, and gives the watch its story.

❹ The watch movement is kinetic sculpture under glass. The eyes are immediately drawn to the mechanism: gears, springs, cams—all driven by an exposed battery. The number of components was dramatically reduced in the Swatch family to approximately fifty, about half the typical number found in analog watches and one of the reasons the watch could be offered at such a low price. Time telling is supported by an outer ring that is a bit too liberal with its markings. Better to have omitted the minute lines, or further differentiated the hour and minute lines à la Jacobsen's City Hall clock. The outer ring also bears an appropriately understated Swatch logo, subordinated to the lower middle position. The logo is sufficiently subtle so as not to interrupt the form or function of the watch, but findable for those seeking the brand. The hour and minute hands are rectangular and playfully colored—a nod to Mondrian?

❺ Prior to Swatch, people typically would own one watch. By the positioning of Swatch watches as fashion accessories first and timepieces second, watch-buying behaviors mutated into fashion accessory-buying behaviors. As with jewelry and shoes, consumers purchased different watches to match different fashion ensembles, in many cases wearing multiple watches at once. While the basic engineering and manufacturing remained constant, well over a hundred different styles were introduced. Swatch did not invent emotional design, but one would be hard pressed to find a better example of how emotional design can impact the success of a product, organization, industry, and country.

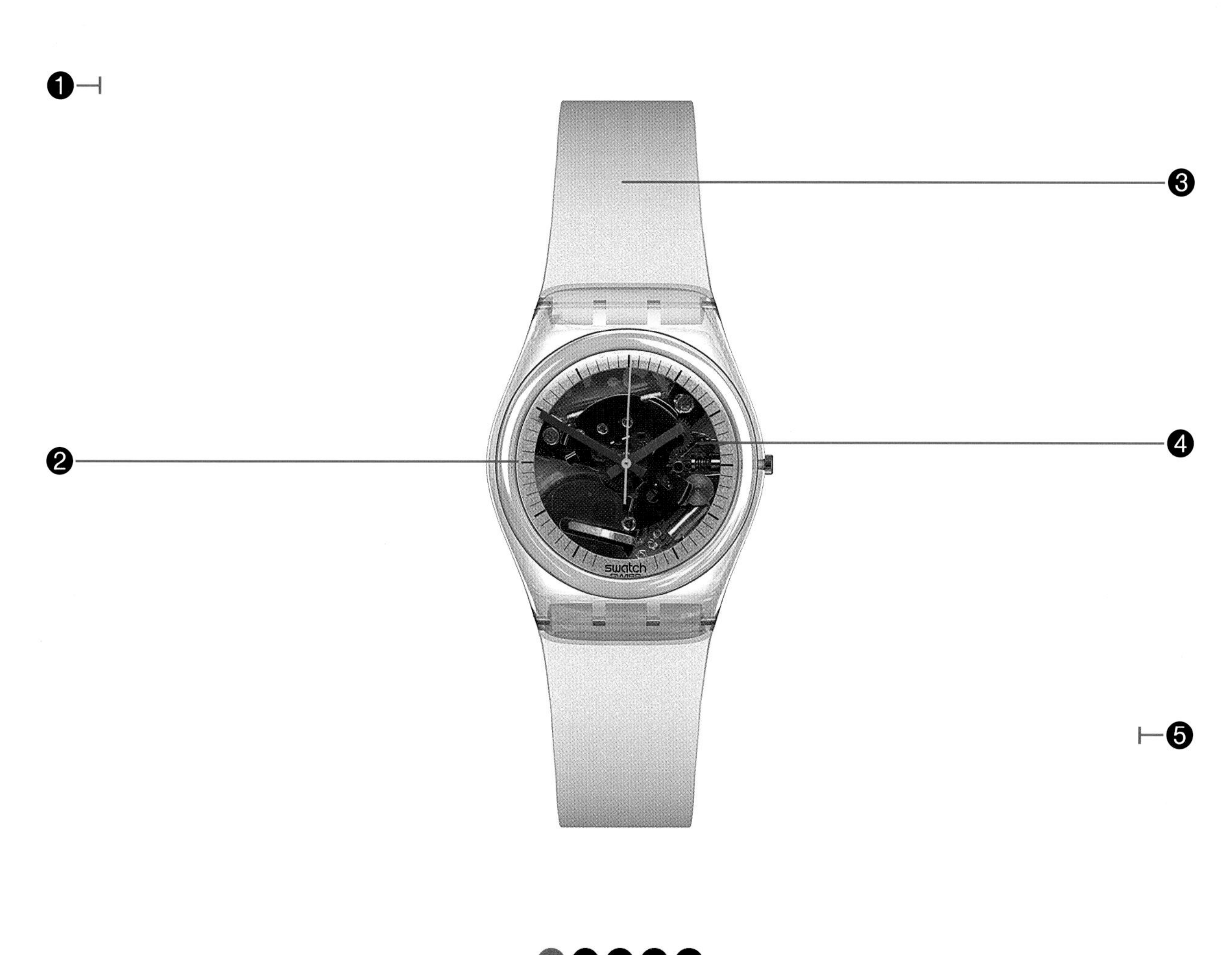

1
2
3
4
5
swatch

Joystick CX40

Atari, 1977

1. The joystick has proven to be an effective interface for translating gross human movements into machine movements. A primary control device for aircraft since the early 1900s, the technology was first adapted for video games in the late 1960s in the Magnavox Odyssey. However, it was not until the introduction of the Atari 2600, equipped with its two CX40 joysticks and popular arcade-style games, that video gaming really took off. The controller is about as simple as a controller can be—a stick mounted to a square with a button—but it is this simplicity that may well have been the key factor in growing a nascent and struggling video game industry. Nolan Bushnell, founder of Atari, comments: “In the beginning, all games were casual. A one-button joystick is something that easily resonates with everyone. Even in the coin-operated space with *Pac-Man* and *Pong*, these games were all very, very simple. We had a much broader demographic and a tremendous number of businessmen who came in during lunch and things that just don’t happen today. I see a striking parallel between the simpler classic times and this casual gaming space, which is really reaching out to those consumers who were disenfranchised by the forty-seven-button controllers…”

2. The form is simple and modern. Although lightweight, the rubber enshrouded stick and solid plastic base feel rugged and durable. The greatest genius of the design may very well be the constraints it placed on game developers—the simplicity of the controller dictated simplicity of game play.

3. Shrouded in a black rubber boot that blends with the case, the stick is straight and symmetrical, enabling users to achieve and adjust their grip from a variety of angles—an early example of good universal design. The stick’s hexagonal shape provides large flat edges that offer tactile directional cues, but the number of faces is at odds with its support for eight directions of movement—an octagonal shape would have promoted better directional mapping. The stick has a short, tight throw, making control movements simple and predictable, and self-centers when not engaged, returning users to a safe neutral position. An accordion base anchors the rubber boot—similar to the gearshift boot found in sports cars—and clearly communicates how the stick is to be used, both functionally and affectively. The dashed line surrounding the circumference of the boot offers some orientation information, but is mostly ornamental, as it suggests a level of directional support not available. Better to have marked the eight supported directions and omitted the dashes.

4. The button is large and red, modeled after the “fire” buttons found on fighter aircraft joysticks. As the sole button on the control, a fact reinforced by the high contrast red-on-black coding, the button invites pressing—which is a good thing, because as the only button, it gets pressed a lot. Unlike the stick, the button positioning lacks universal design consideration, supporting right-handed users only, effectively ignoring ten percent of the population.

5. The base is square and easily held. The simple shape offers key orientation cues for the controller, with the corners indicating the diagonal directions and the edges indicating the up-down and left-right directions.

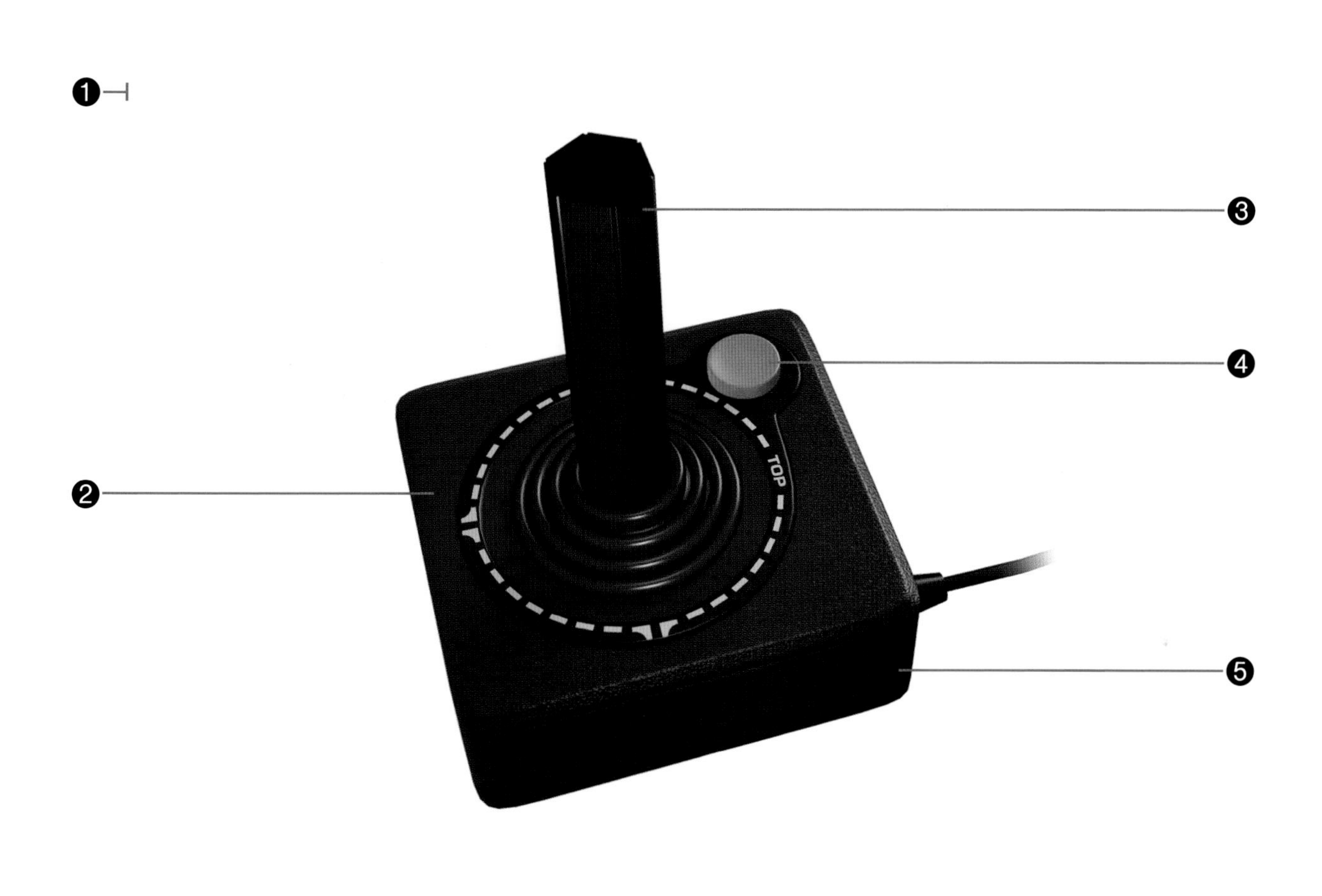
TOP

Juicy Salif

Philippe Starck for Alessi, 1990

❶ The Juicy Salif—the name apparently derived from the French word for saliva, *salive*—has become iconic in the field of product design. Designed by Philippe Starck, the lemon juicer's commercial success is rivaled only by its unapologetic emphasis on form over function. Philippe Starck's reply to such criticism: "Sometimes you must choose why you design—in this case not to squeeze lemons, even though as a lemon squeezer it works. Sometimes you need some more humble service: on a certain night, the young couple, just married, invites the parents of the groom to dinner, and the groom and his father go to watch football on the TV. And for the first time the mother of the groom and the young bride are in the kitchen and there is a sort of malaise—this squeezer is made to start the conversation."

❷ The creature-like form exhibits both Phillip K. Dick and calamari influences (the design was inspired while eating squid). Its symmetry and uniformity combine with its abstractness to stimulate comparisons and analogs—to the unacquainted, it looks like many things long before it looks like a juicer. Its polished cast-aluminum structure achieves a seamless modern aesthetic reminiscent of the earliest conceptions of space ships and alien robots, yet it also appears organic, even sexual, and this tension may be the primary source of its appeal. The product is also strikingly unique, differentiated not only from other juicers, but also from virtually every other kind of product one might find in a normal home. Perhaps the juicer was never intended as such, but rather as a piece of pure sculpture—its purported function a ruse, a Trojan horse employed to invade the kitchens of unsuspecting designerati with modern art.

❸ The body is an inverted teardrop configuration with fluted walls. The flutes converge at the top to create a juicing surface, channeling the juice and pips downward to a sharp point at the bottom, which serves as the exit to the glass. This is the theory. In practice, the smooth flutes do little by themselves, requiring the user to exert considerable energy to extract the juice. The downward force applied makes the long thin legs waver and shift on the counter, requiring one or more of the legs to be held firmly with one hand while the other goes about twisting. This invariably results in the hand interrupting juice flow, dribbling juice on the hand, tabletop, floor, and if lucky, into the juice glass. By this loose standard, just about any object can be called a juicer—but no other object will be as interesting when not juicing, which is, of course, the point.

❹ Three tapering legs extend downward with a slight outward bow, permitting entry and egress for a small juice glass. Knees jut upward from the body at approximately the waist position, promoting the appearance of organic form and minimizing interruption of juice flow. Given their long length and relatively slight cross sections, the legs suffer nominal stability and have been known to break if dropped. However, they are also the element of the design that gives the juicer its striking aesthetic. Freudians will see an abstractly sexual creature, legs spread in seductive pose. The less Freudian, by contrast, may interpret the alien creature to be crouching, preparing to relieve itself into their juice glass. Either way, inescapably, the form factor provokes reaction.

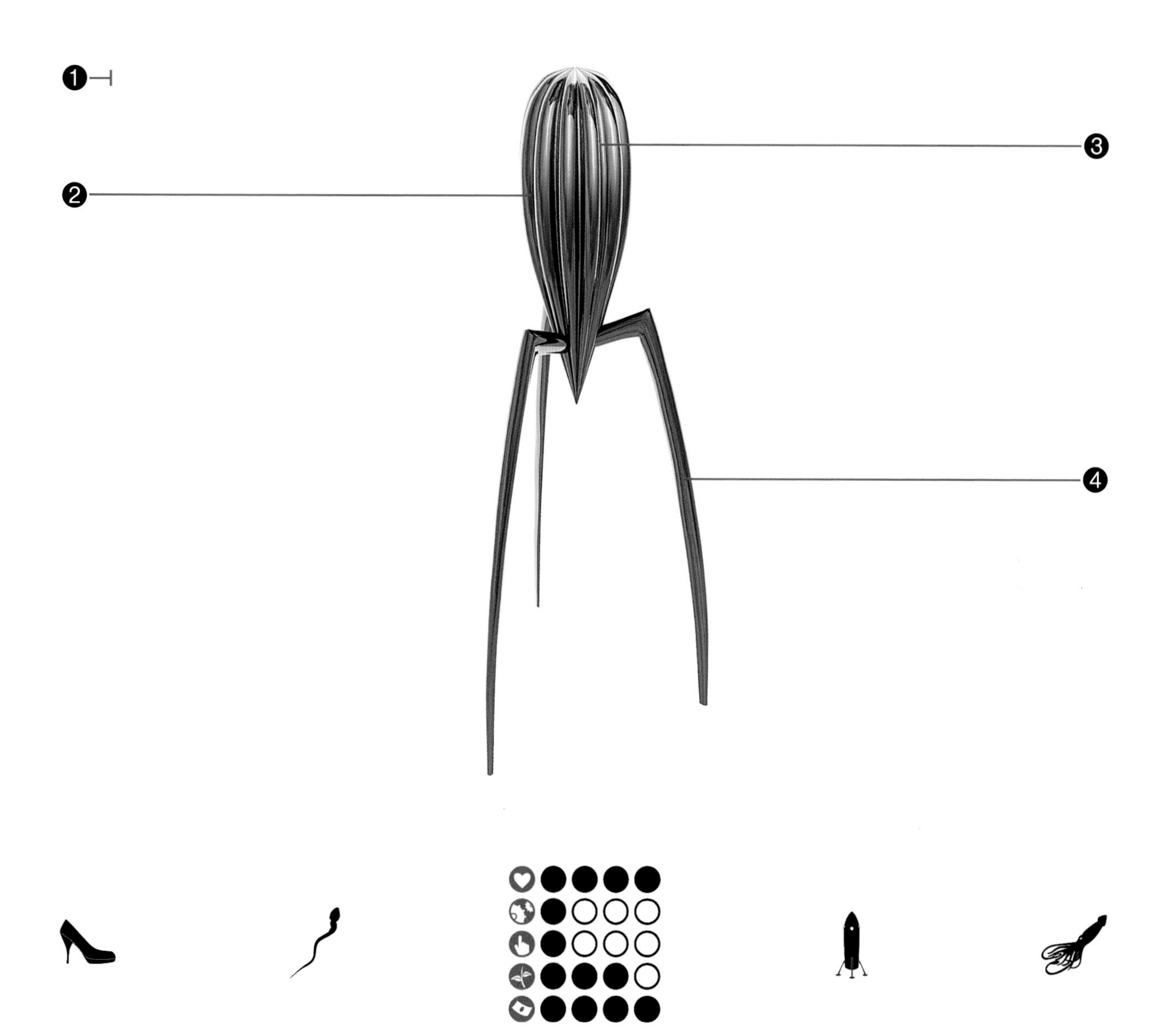

Kryptonite-4 Bicycle Lock

Michael Zane and Stanley Kaplan, 1976

❶ In 1971, bicycle mechanic Stanley Kaplan developed a new kind of bicycle lock that utilized a rigid U-shape configuration versus a flexible chain. He named his lock "Kryptonite," after the fictional element that defeats Superman. Michael Zane read about Kaplan's U-lock in the newspaper, contacted him, and a partnership was soon formed. In 1972, Zane purchased the design and name from Kaplan and founded Kryptonite Corporation with $1,500 from his personal savings. Zane refined the design, and his father, who owned a sheet metal business, assisted with fabrication and manufacturing. With no money for marketing or promotion, Zane staged a promotional experiment in Greenwich Village, attaching a bicycle to a signpost with a Kryptonite lock. After a month, the bike was completely stripped, but the lock and frame remained intact. Publicity from the experiment, in conjunction with a $500 insurance policy for buyers who have a bike stolen, helped make Kryptonite the premium brand in bicycle security for the next twenty-five years.

❷ The form looks like the elongated shackle of a padlock. Its rigid structure is a departure from the chain-based devices of the day, which could be easily defeated with bolt cutters or a hacksaw. In fact, a common observation at the time was that the total weight of a bicycle plus its security chain was always constant, as more expensive, lighter bikes would require correspondingly heavier security chains. The Kryptonite-4 is more manageable and much lighter than a traditional lock and chain equivalent. The body is zinc and nickel-plated steel with a black vinyl coating. Lettering on the drum-shaped barrel is minimal, noting only the name and patent number in high-contrast white. It looks like the Maglite of bicycle locks.

❸ Like a good magician, the lock achieves much through misdirection. The Kryptonite-4 is comprised of two pieces: barrel lock and shackle. An armored tubular shroud conceals and protects the barrel lock, so a thief's attention naturally turns to the shackle. The shackle looks like a big handle. The clear initial affordance is to try to pull on it, detaching it from the barrel. No luck? Next best affordance is to cut or spread the long, thin arms of the shackle, the apparent weakest sections of the lock. Accordingly, early approaches to defeat the lock used hacksaws to cut the shackle, or large poles to spread and separate the arms. These methods were nontrivial, nondiscreet, typically nonsuccessful—and, most importantly, nonsurprising. Short of a hydraulic jack or acetylene torch, the shackle of the lock is impervious to attack through commonly available means. Unfortunately, the engineers at Kryptonite may, too, have been overly fixated on the shackle's tamper-with-me-first affordances. Consistent with the maxim that "a chain is only as strong as its weakest link," it turns out that the barrel locks of many Kryptonite models could be easily picked, using the plastic barrel of Bic pen—remove the innards from the pen, twist the barrel into the lock, jiggle it around and, with practice, the lock would release in a matter of seconds. The locks were updated to defeat this method of attack, but the episode underscores three lessons: 1. Misdirection can be an effective tool in the design of security systems; 2. Product designers and engineers are as susceptible to their own misdirection; and 3. In the age of the Internet, one must assume that methods of defeating security systems, once discovered, will be immediately and widely known.

KRYPTONITE
Tough World, Tough Locks

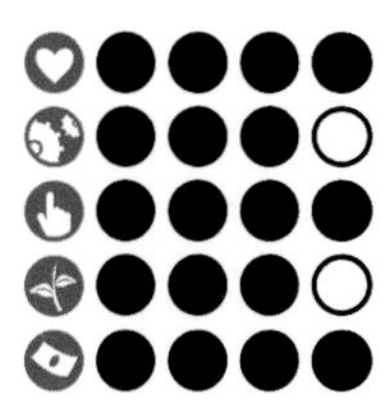

LC4 Chaise Longue

Le Corbusier, Pierre Jeanneret, and Charlotte Perriand, 1928

❶ A modernist interpretation of the bentwood chairs and chaises of Michael Thonet, the LC4 exemplified the new minimalist possibilities afforded by tubular metal structures. The chaise, dubbed the "relaxing machine" by Le Corbusier, found inspiration from resting soldiers, and was designed to mirror the body's natural curves. Charlotte Perriand comments: "We had prepared sketches showing the common ways of sitting in the West. Then we started looking for structures. Giedion says that when you sit down it is as if you are losing the power of your muscles. That is why it is necessary for the skeleton to 'support'…We then added canvas and fastening attachment to the frame, completing them with the help of horsehair, springs, and other materials. What came out of all this was a fairly simple system: the Chaise Longue, proceeding from the basic idea of a simple soldier, who, when he is tired, lies down on his back, puts his feet up against a tree, with his knapsack under his head. This medically recommended position, with legs raised, is a very comfortable one. And that's how we came up with the idea of the Chaise Longue."

❷ The form is abstractly anthropomorphic, its supporting structure largely concealed except for a long and gentle arc support. The arc struts sit atop rubber-covered cross-members in the base, holding the chaise firmly in position at any angle of inclination through force of friction. The effect is a reclined figure floating in air, and makes one forgive the otherwise uninviting thin and narrow laying surface.

❸ The fully reclined orientation of the chaise is known as the zero gravity or 90/90 position, essentially the catenary position for the human form—one wonders if Le Corbusier was playing with the concept of an anthropometric catenary supported by a geometric catenary. At full recline with the hips and knees bent, the back assumes the neutral position, maintaining the proper S-curve of the spine and relieving pressure on the back. Additional benefits of the position include improved circulation, reduced stress on the heart, and facilitated breathing. There is a reason astronauts assume this position on liftoff. The headrest looks like the head of an abstract figure, and its tubular shape fits with the geometry of the chaise. Tubes, however, do not make the best headrests, and the trade is form for comfort.

❹ The key innovation is the bow-shaped frame, which gives the chaise its distinctive geometry, dramatic levitating effect, and ease of adjustability. The arc support is in the shape of a catenary; an inverted arch formation that is in uniform equilibrium with gravity. The chaise is adjusted by lifting and repositioning on its rectilinear base. Lying positions range from near upright to full recline.

❺ The H-shaped base is stout and functional, but relative to the curvy elegance of the chaise, pedestrian in form. An attempt of sorts was made to streamline the elements, fashioning the legs after airplane wings, but this just made a bad situation worse—it looks like a project that ran out of time, and the wings a rushed attempt to make the base look "designed." Better to have kept the design basic and lose the wings, or expand the use of arcs and arches in the structure and composition.

LCW Chair

Charles Eames and Ray Eames, 1946

❶ In what *Time* magazine declared to be "the chair of the century," the LCW chair, for "Low Chair Wood," is the culmination of the Eames's long and trying search for the ideal application—or as they would say, the purely *honest* application—for molded plywood. With wood-molding techniques honed developing molded plywood splints, stretchers, and glider shells during World War II, the Eames turned their attention to furniture design post war. After more than five years of failed attempts to form a chair from a monolithic shell, the LCW was born, having emerged from a kind of Petroskian "form follows failure" process, where the final product resulted more from a seemingly unending succession of failed attempts than anything else. The result is a set of molded plywood components, each optimized for a specific function, all carefully constructed and integrated to address the needs of the material, manufacturer, and user. In the Eames's film, *Design Q&A*, Madame L'Amic asks Charles Eames, "To whom does design address itself: to the greatest number, to the specialist of an enlightened matter, to a privileged social class?" Eames replies, "Design addresses itself to the need."

❷ From reverse profile, the form is biomorphic, doglike, striking a playful pose that looks upward to its master. A single piece of molded plywood acts as the back and neck, connecting the seat, backrest, and legs. The piece dips near the back, providing clearance for the tilting seat, while the elevated front raises the seat up from the frame, appearing to serve it up for the sitter. Legs are reinforced with additional plies for added strength.

❸ The backrest is an isosceles trapezoid, with a wide base and narrow top. From the front, the slope of its sides is parallel to the slope of the legs, aligning the elements and creating harmonious integration. The panel is gently contoured to fully support the back. The backrest flexes firmly, deliberately, a function of the pliability of the material working in concert with rubber shock mounts, and is one of the first examples of a chair to be responsive in this way.

❹ The seat is wide and comfortable, possessing the most elegant and functional geometry of the chair, but its access requires a youthful agility that may not agree with larger bodies or arthritic limbs. The midplane of the seat cuts very close to the "Golden Section" of the chair, creating an aesthetic proportion. The center of the seat dips near the back and gently wings upward at the sides, securely cradling the user but making exit more of a challenge. The front edge is a waterfall configuration, which avoids snags on entry and reduces problems of circulation under the thighs.

❺ The chair's simple form, uniform wooden construction, and low posture will lead nondesigners to conclude that it is a chair for children, superficially resembling the small wooden chairs found in many elementary schools. This association would likely not offend the Eameses, as the goal of their design was to make an affordable, comfortable chair that could be easily mass produced. However, they might be offended that it is not their chair in the classrooms, but low-priced knockoffs, as modern versions of their chair are priced and marketed strictly to affluent designerati, not the masses.

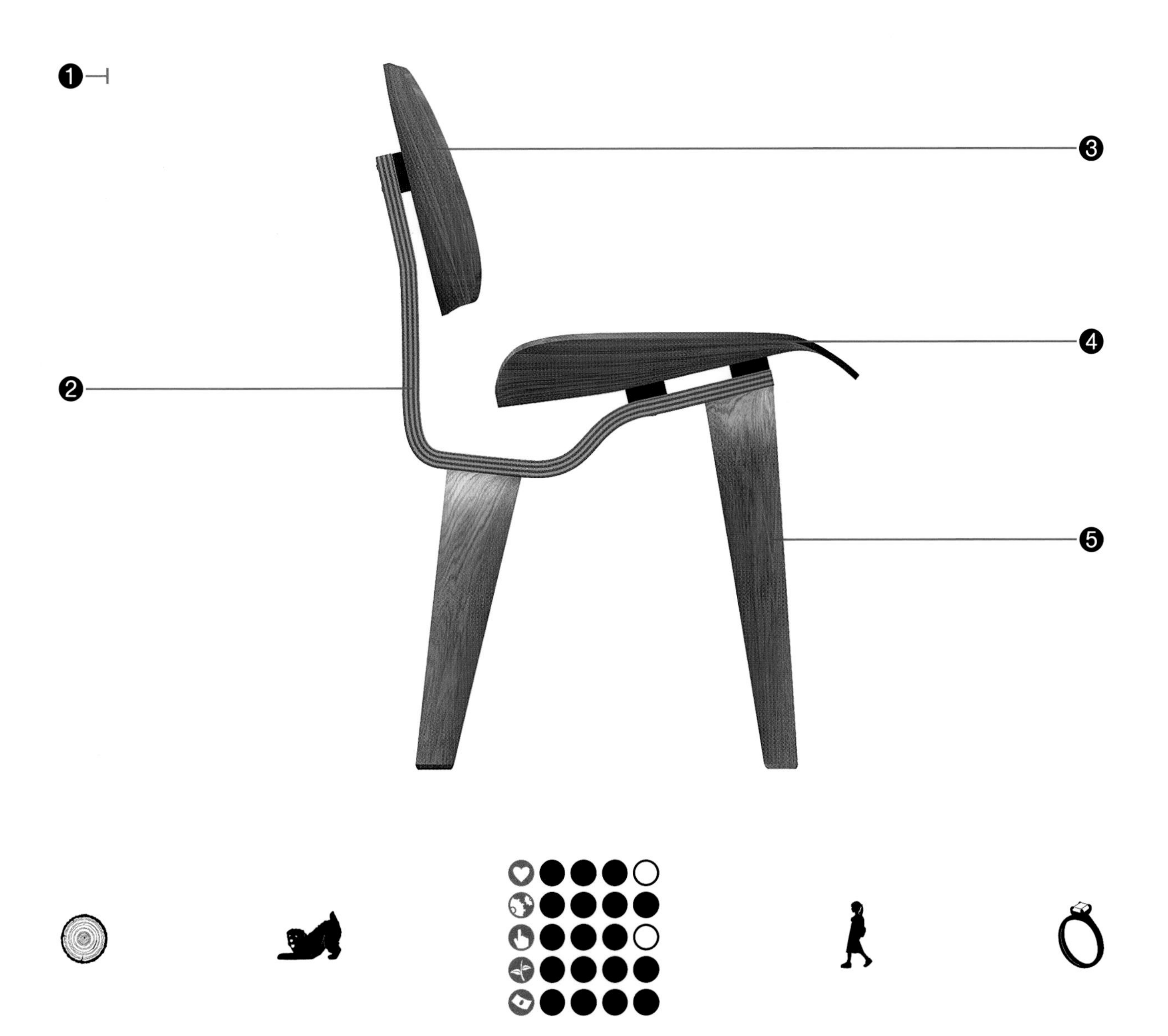
1
2
3
4
5

LED Watch PH-1055

Philippe Starck for Fossil, 2004

❶ A modern interpretation of the 1970s-era LED watches, the PH-1055 carries all of the signature stylings one would expect from Philippe Starck: minimalism, wit, irreverence, and an aesthetic more about starting a conversation than telling time. It is the quirky minimalism of the LED watch that first grabs the attention here—is it a bracelet, or some kind of Braille watch? The answer does not really matter as long as the question is asked. Philippe Starck comments: “Why make complicated, when you can make it easy? But I can tell you, to make something simple is ten times more difficult than to make something complicated. To fight, millimeter by millimeter, to have less. To fight the *component électronique* to [make it] more simple. Products are so stupidly complicated, just to show the intelligence of the people who do it. They don't show the intelligence of the people who design it. They show the stupidity of the people who design it.”

❷ The watch defines *geek chic*. The LED display references the earliest digital devices, while the body appears transported to 2004 from some future time. Its design is gender neutral, though men will likely find the retro design and rectangular display more appealing than women. Perhaps due to its black urethane body, the timepiece passes for a diver's watch, and boasts water resistance to fifty meters. The precision with which the parts are fitted is extraordinary, especially considering the small size and complexity of the light pipes. Accordingly, the watch exudes quality and fine craftsmanship. When activated, the pipes illuminate bright orange, contrasting nicely with the black body. Legibility is less than optimal, but the display design isn't about optimality—it is about getting attention.

❸ Unlike traditional watches, the default display mode for the PH-1055 is “off”—that is, the user must press a button to see what time it is. In terms of usability and basic efficiency, this is clearly an unnecessary step. In terms of aesthetics, however, it is consistent with Starck's philosophy of designing products that refuse to be ignored or casually acknowledged. The watch wants explicit and undivided attention each time it is used, and it holds time hostage until the user is willing to give it what it wants. Once pressed, light emitting diodes illuminate the light pipes in bright orange and reveal the time in a vertical configuration—hours on top, minutes on bottom—but only for a few seconds, and then the watch again sleeps. A second button press is required to learn the date. This coyness aside, a practical benefit of both the LED and the brief reveal is minimal power consumption. A thirteen-segment display is used, a variant of the more common seven-segment displays used in traditional LED and LCD displays. The addition of segments permits longer and more rectangular numerals, but the cost is visual complexity.

❹ The strap is a work of art—a single fluidic piece of polyurethane. The light pipes individually extrude through the material, and the Starck logo is blind debossed at the bottom. The contrast between the LED display and monolithic strap gives the watch its future-retro aesthetic. The strap is locked into place using a stainless-steel butterfly buckle, also bearing a blind deboss of the Starck logo—in case the user didn't notice the logo on the face. Lose the superfluous buckle logo, and the design stands as the digital equivalent of the classic Museum watch.

1
2
3
4
STARCK

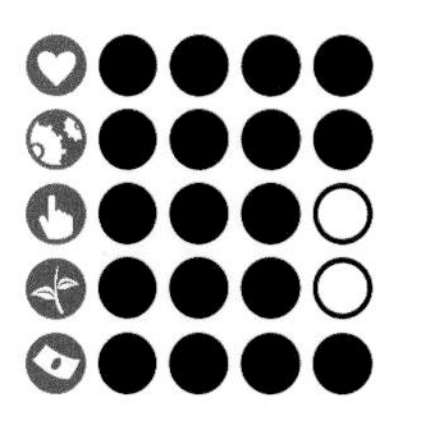

Lil' Pup Dog Bowl

Mark Kimbrough and Pearce Jones for Wetnoz, 2001

1. Perhaps the most valuable and least understood skill in product design is opportunity recognition—that is, recognizing when problems can (and cannot) be materially impacted through design. Although difficult to formalize, one constant among those proficient at the skill seems to be the ability to achieve a deep understanding of user needs, needs that in many cases users themselves are unable to recognize or articulate. Such is the case with the founders of Design Edge and Wetnoz, both of whom have demonstrated an uncanny knack for not only recognizing opportunities where design can make an impact, but also capitalizing on them—as they have with their ultramodern line of dog bowls. Mark Kimbrough comments: "[After eight months of market research] we discovered the pet industry, a $30 billion industry in the U.S., had no recognizable brands when it comes to hard goods. Who makes your dog bowl? The shape of a dog bowl is fairly iconic in our minds. It's round with slightly slanted sides. Well, who made that rule? ... Everyone is [now] scrambling to do what we do, which is find market segments that have not been infiltrated by contemporary design. And once we get Wetnoz on an even keel, we intend to do it again."

2. The form is simple and modern, reminiscent of Nambé serving containers. The one-piece body of the bowl is constructed of surgical-grade stainless steel, which conveys high quality and complements the high-end stainless steel appliances and cookware found in contemporary kitchens. The absence of seams offers no sanctuary for bacteria and makes cleaning a simple operation. The handle and base are accented with black polymer, to soften the steel edges and provide grippy surfaces. The design astutely acknowledges two user groups: owners and dogs.

3. The dipping cut of the bowl and the oval cutout handle make both resemble large mouths, agape, impatiently waiting to be fed—subtle ironies infused through form factor alone. One wonders to what degree, if any, these clever nuances influence subconscious impressions, even if they are not consciously recognized. Regardless, they are a joy for those who get the joke. Perhaps this is one reason the bowl is increasingly adapted for human versus canine use—reports suggest that the bowls are being used for everything from doling out finger foods in buffet lines to acting as trendy containers for car keys and related miscellanea. The low-slung front defines the access point for the dog, while the high back keeps the food in the bowl. The shape of the bowl with the support of the cutout handle enables it to be used as a scoop for kibble, avoiding dirtying the hands of delicate dog owners. Considering the high price point of the bowl, users deserve to have their hands unsullied.

4. The base sits naturally on a circular flat supported by a one-piece rubber backstop. This makes the bowl slightly front-heavy, resisting scooting and tipping backward when accosted by jabbing muzzles. As with the bowl itself, the backstop, too, has a sense of humor, abstractly shaped in the form of a dog bone. It is important to note than no function or usability is compromised in incorporating these design wits—the backstop fits the bowl naturally and serves its purpose well. The rubber components are easily detached for cleaning. An elevated bowl stand is available to minimize stress on joints and muscles. A welcome accessory for those with older dogs.

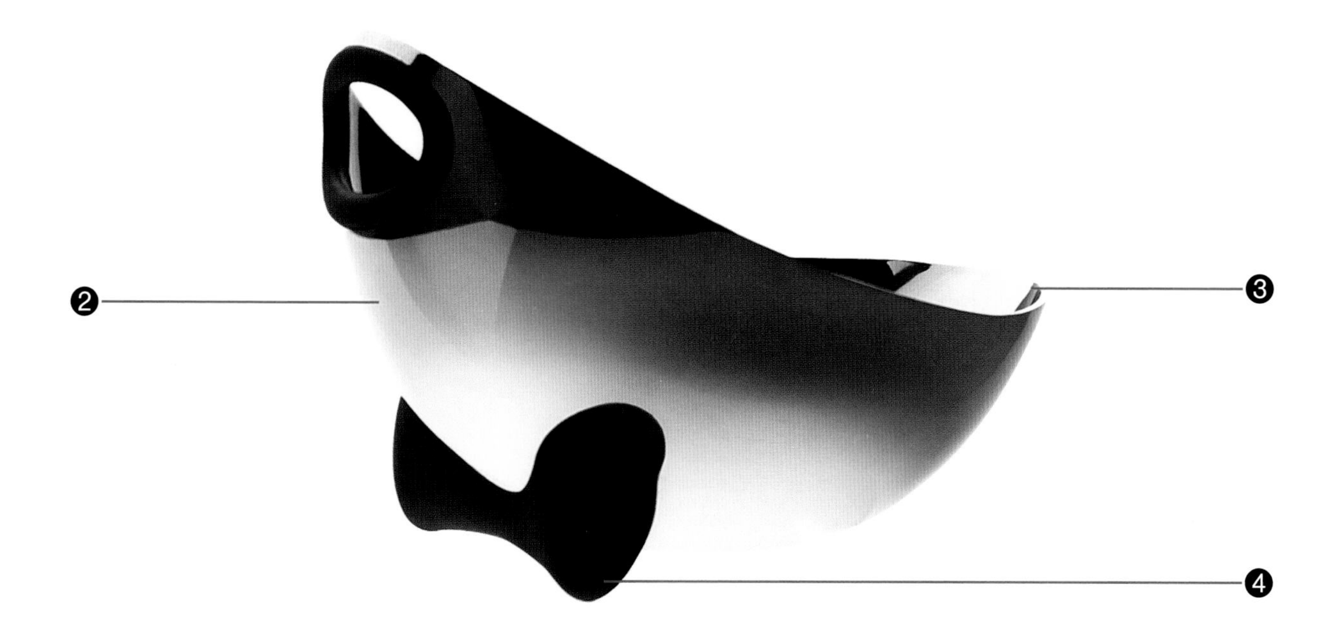

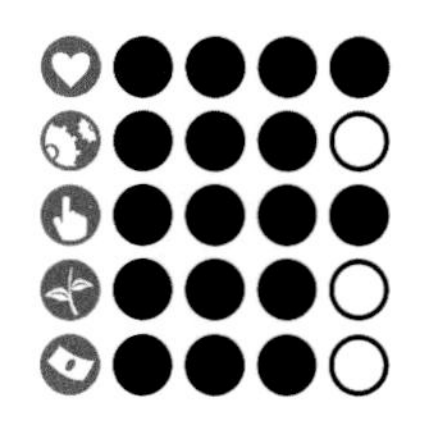

Lounge Chair and Ottoman

Charles Eames and Ray Eames, 1956

1. Encouraged by friend and noted director Billy Wilder to build an "ultra, ultra, ultra comfortable modern lounge chair," Charles and Ray Eames responded with a design that was at once modern and traditional—an interpretation of the traditional club chair, luxuriously adorned with rosewood veneer and leather upholstery. The chair represented a new kind of matter-of-fact modernism that appealed to both designerati and more general, albeit affluent, audiences. The design exemplifies the Eames's belief that great design emerges from a mandatory process of evolution, beginning with the simple, graduating to the esoteric, and then graduating back to the simple, a philosophy Charles Eames often summarized using the "banana leaf parable": "Under the caste system of India the lowest form of caste would eat off a banana leaf. The next highest would eat off a ceramic dish. The next highest had a glazed dish. Then perhaps a brass dish, a bell-bronze one, then a plated one, then maybe a gold one. But then, when you got to the very highest, they ate off a banana leaf again."

2. The articulated molded plywood shells are the easiest element to like in the chair. They are sculptural, functional, organic—the bases of the chair and ottoman look like stingrays. Their gentle curves and overall symmetry make the shells look unwoodlike, leading the unacquainted to wonder by what magical process could rosewood be bent in this way. The U-shaped shells cradle the cushions, providing uniform support and minimizing user contact with anything other than the cushions.

3. The form is aptly summarized by Charles Eames, who aspired the design to have the "warm, receptive look of a well-used first baseman's mitt." Unfortunately, it also has the questionable visual appeal of a first baseman's mitt, an aesthetic deficit that did not go unrecognized by the Eames. Charles once commented that there was a "sort of ugliness" to the chair, and Ray described early versions of the chair as "un-designy." Comfortable, no doubt, but subtract the Eames' brand and "Veblen effects," and what remains is a well-marketed, visually unresolved lounge.

4. The cushions are plush, portly, and ostentatious. The black leather upholstery has definite bourgeois appeal, possessing an executive aesthetic that appeals to a generation mesmerized by fins on cars. The cushions were originally filled with a combination of down, duck feathers, and foam, making them feel as comfortable as they look. Coupled with its well-integrated ottoman, the lounge anticipated the television-watching, couch-potato posture that would soon become universal in American households.

5. The five-point aluminum base of the chair permits 360-degree rotation, with the odd number of supporting arms allowing the chair to be in any position without looking off-center. The four-point base of the ottoman does not pivot. It is the metallic elements of the design—bases, arm rests, and back connectors—that compromise its integrity more than any other, appearing *ad hoc* and at odds with the sensuousness of the leather and wood. If only there had been one more iteration in the Eames's design process, one in which cleverly molded plywood forms replaced these disjunctive metallic components, then perhaps the lounge chair could live up to its now incorruptible iconic status.

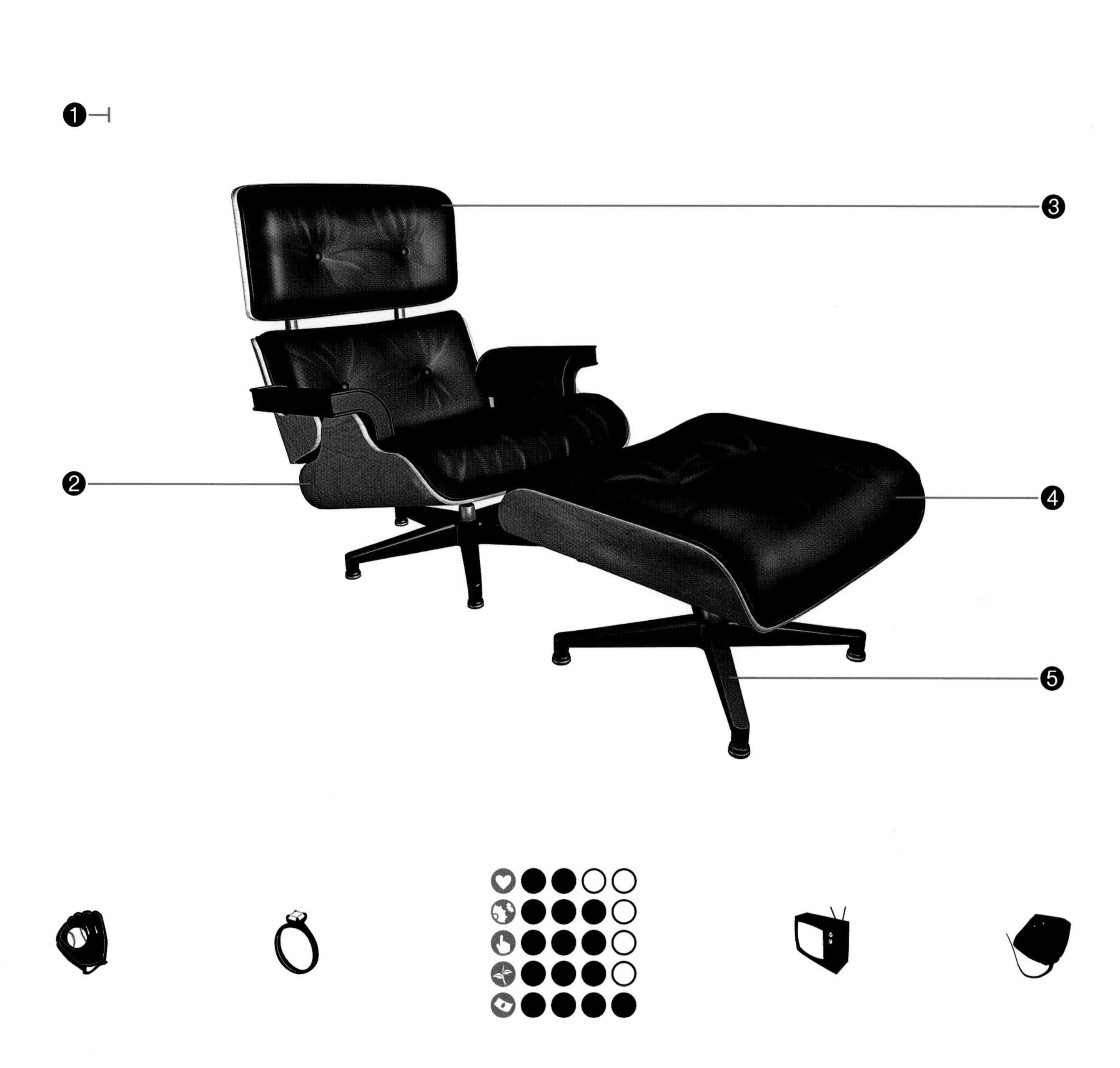
1
2
3
4
5

Macintosh

Apple, 1983

1. The Macintosh introduced the possibilities of industrial and user-centered design to the computing masses, emphasizing usability and user experience over cryptic and often inaccessible hardware and software capabilities. The result is as much art object as computer, and like a work of art bears the signature of the original team members on its inner case. Where does such a revolutionary design find its inspiration? Often from other revolutionary designs. Andy Hertzfeld, original Macintosh design team member, recounts one design discussion: "I heard loud voices emanating from Bud [Tribble's] office, which was adjacent to mine, apparently engaged in a spirited discussion. 'It's got to be different, different from everything else… We need it to have a classic look that won't go out of style, like the Volkswagen Beetle,' I heard Steve [Jobs] tell James [Ferris]. 'No, that's not right,' James replied. 'The lines should be voluptuous, like a Ferrari.' 'Not a Ferrari, that's not right, either,' Steve responded, apparently excited by the car comparison. 'It should be more like a Porsche!'"

2. The inspiration may have been an exotic car, but the computer's *prima facie* appeal has more to do with its baby-face appearance than its Porsche-inspired lines: vertical versus horizontal orientation, chamfered top edge as high forehead, inset monitor as large eyes, disk drive slot as mouth, and a small recess at the bottom as a small chin. If the baby-face connection is not immediately obvious to some users, it becomes obvious when they are presented with an iconic version of the computer's face, smiling, at startup. Note that baby-face features convey a light, playful quality, which more serious users (e.g., business professionals) will not (and did not) find appealing.

3. The edges of the face are chamfered, softening the rectangle and directing the eyes inward. The distance between the top of the case and the display bezel is small, consistent with the sides, and the inset of the bezel itself is shallow, avoiding what Jobs referred to as the "Cro-Magnon forehead" possessed by other models. A carrying handle is cut into the top, enabling the computer to be easily carried, even playfully swung, with one hand. Practical and functional, but also a bit toylike—could a serious computing device really be carried in such a manner?

4. Standing on the shoulders of the work at Stanford Research Institute and Xerox PARC, the Macintosh was the first commercially successful product to employ a graphical user interface (GUI). The Mac GUI—desktop metaphor, fixed menu system, direct manipulation icons—revolutionized human-computer interaction, and remains fundamentally unchanged to this day.

5. The disk drive is easily accessible from the front. Centering versus right-aligning the drive slot would better accommodate both left- and right-handed users, achieve better visual balance with the left-aligned Apple logo (the grip-well serving as the visual counterweight), and appear more mouthlike, strengthening the baby-face proportions of the design.

6. The design language of the computer is consistently applied to the keyboard and mouse, both of which share the same ABS material, chamfered corners, and minimalist aesthetic. The keyboard is more akin to a typewriter than a computer keyboard, a simpler design lacking the numerous control and function keys that so often clutter computer keyboards. The basic proportions of the mouse and the face of the Macintosh are equivalent, with the mouse's single button approximating the position of the screen.

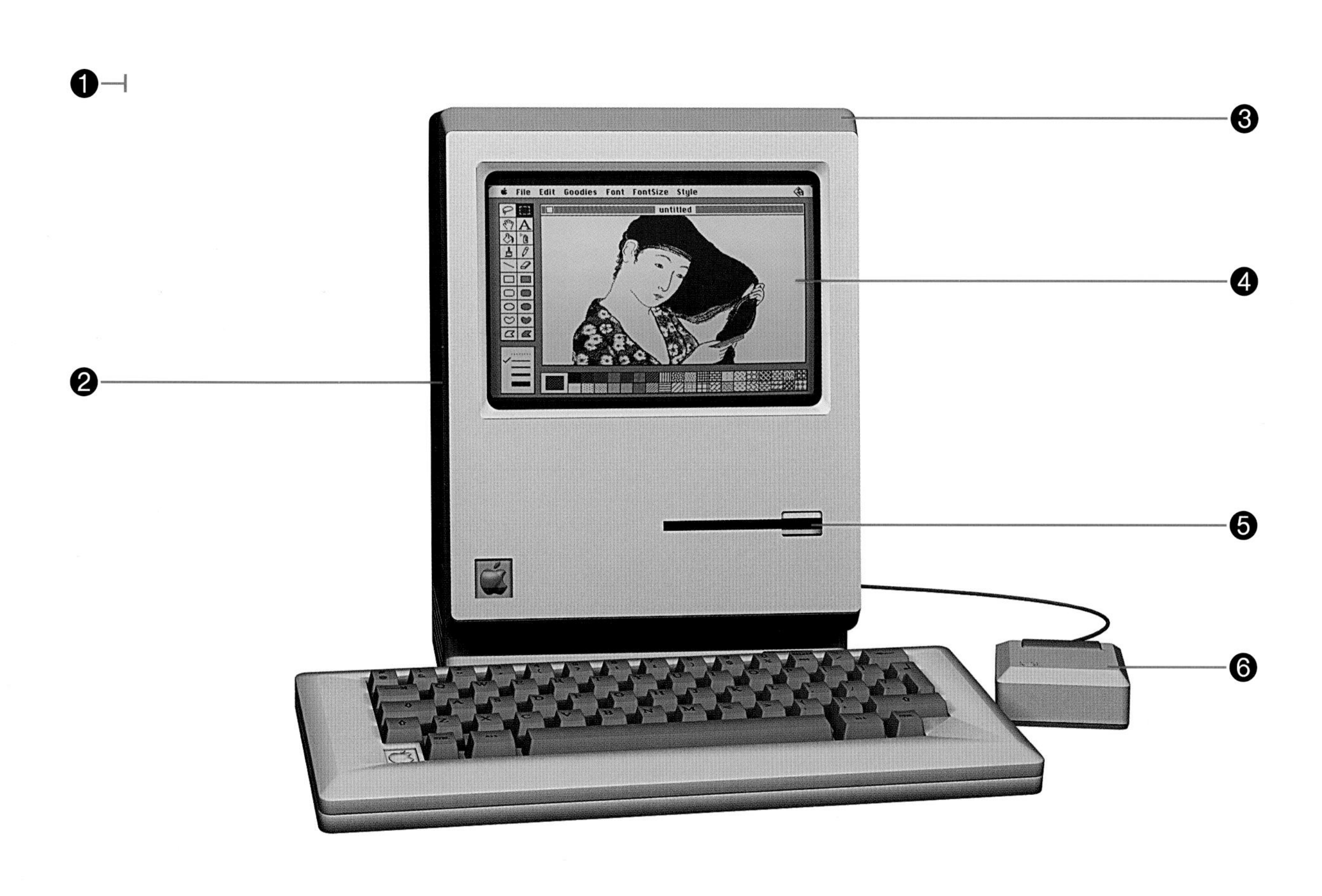
1
2
3
4
5
6
File Edit Goodies Font FontSize Style
untitled

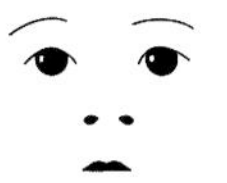

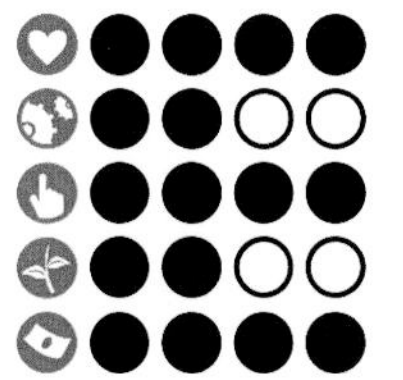

PORSCHE

Maglite Flashlight

Anthony Maglica, 1979

1. Like many great products, the success of Maglite regards understanding and solving deep problems of product design — problems, in many cases, users are not even able to articulate. The flashlight was originally designed for the public safety sector, but it quickly established itself as a viable premium consumer flashlight due primarily to its reputation for ruggedness and reliability. Anthony Maglica comments: "I knew someone in the 1970s named Don Keller, who later came to work for me, and still does. He had been a policeman, and he knew cops who had problems with their flashlights. The flashlights were flimsy, made of plastic, and if you dropped them, they would break. Keller said that if someone could make these out of metal, they wouldn't break ... I designed one with a pushbutton switch instead of a slide switch, and an adjustable beam, so you could go from flood to spotlight ... We get letters from customers all the time, with stories about the lights and what they've been through. Like someone losing a light in a lake and finding it the next year, and it still worked."

2. The flashlight is long, cold, and heavy, the result of its machined aluminum construction combined with the ample batteries required to power the bulb. The aluminum is anodized inside and out to resist corrosion, and the three seams are sealed with O-rings for water resistance. The flashlight is black, reinforcing its identity as a serious tool and premium product. Maglites are often associated with clubbing and abuse by law enforcement, and holding the flashlight does little to undermine this association — it looks and feels like a weapon.

3. The gripping surface is machined into the aluminum as diamond knurls, which wrap the full circumference of the body. This textured section tells the user where to hold the flashlight, which happens to also be its center of gravity. The knurls provide an effective gripping surface that do not wear or significantly degrade in adverse conditions. Batteries are exchanged through a tail cap, wherein a spare lamp, protected by a foam lamp protector, is included in case of damage or failure of the primary. Such a simple thing, the spare lamp — more important than the redundancy it provides is the message it communicates to users: this is a serious product; it will work when you need it to; it is worth whatever you pay for it.

4. The power switch is black rubber and resides within a thumb-shaped well in the metal. The switch blends with the body, but is sufficiently distinct in finish and texture to be clearly visible — it invites pressing, soft and warm relative to the surrounding metal case. The rubber switch also offers a level of water resistance and durability, and it self-cleans by gently rubbing the internal electrical contacts when activated. Three switch positions are supported: on, off, and signal, which flashes intermittently. It may be the perfect button.

5. The head is tulip-like and just larger in diameter than the body. Twisting the head alternates the light beam between spot and flood, a unique innovation at the time of its introduction. Bulb strength and quality are excellent. The only branding resides on the circumference of the head as white lettering, subtle in overall effect, but clearly visible and effective at instantly branding the flashlight.

1
2
3
4
5
MAG-LITE

Model 302-F1 Telephone

Henry Dreyfuss Associates for Bell Laboratories, 1938

❶ Often called the "Lucy phone" due to its use by the lead characters of the *I Love Lucy* show, the Model 302 is perhaps the first phone that was "designed" in something approaching the modern sense of the word. Then, however, as is often the case today, there remains confusion between "art" and "design." Henry Dreyfuss comments: "In 1930, shortly after I opened my office, a representative of the Bell Telephone Laboratories called on me... His mission was to inform me that the Bell Laboratories were offering $1,000 awards to each of ten artists and craftsmen for their conceptions of the future appearance of the telephone. It was flattering to be included in such a group, and the prospect of a thousand dollars was attractive. But I suggested that a telephone's appearance should be developed from the inside out, not merely created as a mold into which the engineers would eventually squeeze the mechanism, and this would require collaboration with Bell technicians. My visitor disagreed, saying such collaboration would only limit a designer's artistic scope. Several months later he returned with a changed point of view. He admitted frankly that the telephone of the future submitted by the ten commissioned artists had been unsatisfactory... Now he wanted to hear my ideas about design from the inside out."

❷ Originally made out of cast zinc and then later thermoplastic, the Model 302 was near indestructible. Its affect is utilitarian, but the lines of the phone are somewhat stylish. The handset has clear affordances for gripping, orientation, talking, and listening—it is a classic of ergonomic design. The phones were predominately black, but were available in eight other colors near the end of their production run in the mid-1950s.

❸ The F-1 handset is a dramatic improvement over the original E-1, which felt bulky and looked bony. The handset bears a slight arc that both softens the angular features of the phone and, as the high point, indicates that the handset is to be picked up. The large receiver and transmitter offer clear affordances for hearing and speaking, with the attached cord acting as an anchor to indicate proper orientation. Picking the handset up automatically readies the phone for dialing. Setting the handset in the cradle automatically terminates the call. On an affective level, the near symmetric transmitter and receiver look like ears—oversized ears, Mickey Mouse ears—giving the unit an endearing quality.

❹ Long before the iPod Classic, the rotary dial phone was the preeminent rotary input device. To dial a number, insert a finger into one of ten finger holes, rotate the dial clockwise until the finger reaches the stop, and then remove the finger, which allows the dial to return to the resting position. For younger readers, this is to dial just one number. The method is clunky and unforgiving—imagine dialing six numbers to have a finger slip out prematurely while dialing the seventh—but constraints imposed by the one-way clockwise directionality of the dial and finger stop makes the control surprisingly intuitive. The dialing face is appropriately angled for viewing and use from a standing or seated position. The center region permitted a circular paper insert, which was typically used to indicate a phone's number.

❺ The body is solid and heavy, sloping inward and upward like a pyramid, leading the eyes to the handset. The design pays homage to the 1930 Bakelite plastic phone designed by the Norwegian painter Jean Heiberg.

1
2
3
4
5
ABC 2
DEF 3
1
GHI 4
JKL 5
MNO 6
PRS 7
TUV 8
WXY 9
OPERATOR 0

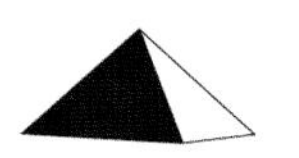

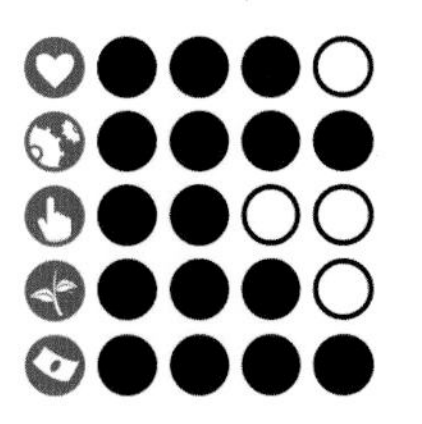

Moka Express

Alfonso Bialetti, 1933

1. The inventor Woody Norris once commented that he invents by analogy, constantly on the lookout for things that work well in one context so that their "tricks" can be applied by analogy in other contexts. This approach is a common one and responsible for a good many product innovations, including the Moka Express. While watching his wife do laundry, Alfonso Bialetti observed the workings of their primitive washing machine: a fire, a bucket, and a lid with a tube coming out of it. The bucket was filled with soapy water, sealed with the lid, and then brought to a boil over the fire, at which point the vaporized soapy water was pushed up through the tube and expelled on to the laundry. Bialetti imagined a similar mechanism for coffee, one in which a lower chamber filled with boiling water would force steam up through coffee grounds and then condense in an upper chamber. Many prototypes later, the Moka Express was born.

2. The basic design of this espresso maker is unchanged to this day: two stacked octagonal chambers that taper inward toward the coffee-containing middle. In profile, it looks like a geometric Art Deco rendering of the female body, which is one reason the form factor has stood the test of time. The maker is made of aluminum, which is said to better maintain heat in a uniform way. The handle is made of Bakelite to minimize heat conduction. Its ergonomic shape assists gripping and its sharp angle keeps the pouring hand clear of the hot metal lid. The famous Bialetti logo was added in the 1950s—a well-dressed man with a moustache, finger raised to order another coffee—a caricature of Alfonso Bialetti himself, created by Paul Campani in 1953.

3. Bialetti has sold over 270 million Moka Expresses globally and can be found in nine out of ten Italian homes. To what can this extraordinary success be attributed? The Fascist Italian government dubbed aluminum the "national metal of Italy" in the 1930s. The metal was promoted as embodying the traditional values of Italian craftsmanship and fine design, and it therefore was a logical choice for Bialetti's new invention. At this time, coffee was consumed predominantly by men in cafés. As such, espresso machines were designed and manufactured for cafés, not consumers—they were large, complex, and expensive. The Moka Express was, in a way, the anti-Starbucks phenomenon, enabling the average Italian to enjoy café-quality espresso at home. With the fall of fascism in the 1940s and the rise of consumerism in the 1950s, the "café in the home" gave women increased access to the previously restricted coffee ritual. Women could not only drink espressos in their home, they could drink espresso prepared by their husband. This reversal of traditional gender roles liberated women, both literally and symbolically, to explore other nondomestic pursuits. Which leads to Alfonso's son, Renato Bialetti, who assumed control of the business in 1946. He engaged in a series of aggressive national advertising campaigns that exploited billboards, newspapers, magazines, radio, and later television. The advertising campaigns were as progressive as aggressive, leveraging the coffee maker's symbolism for women's rights and modernity. So no one variable accounts for the success of the Moka Express. Its success involves the fall of fascism, rise of egalitarianism, power of advertising, and the desire of the average Italian family to be able to make a cup of fine joe in the comfort of their homes.

1
2
3

MoneyMaker Pump

Martin Fisher, Nick Moon, and KickStart, 1991

❶ While the percentage of the world's population living on less than $1 per day halved between 1981–2001, sub-Saharan Africa was the one region that did not participate in this positive trend. The situation in Africa is believed by many to be so extreme that its people are too poor to work their way out of poverty. Many economists believe the only option is to fight this degree of poverty with charity—a lot of charity—on the order of trillions of dollars a year. The people at KickStart have a different theory: sell people, however poor, the tools they need to make money, and then get out of the way. Martin Fisher comments: "Eighty percent of the poor people in Africa are poor rural farmers. They have one asset, a small plot of land, and one basic skill, farming… However, the existing technologies, the petrol pumps and the electric pumps, are simply too expensive for them to buy. And so we have developed a line of human-powered pumps. These small machines, which look like a small take-home StairMaster machine, are operated by walking back and forth, can pull water from a shallow well or a pod or a river, and spray it through a hose pipe and irrigate as much as two acres of land."

❷ The primer blue treadle pump is utilitarian in appearance and construction—to Western eyes it looks and works like exercise equipment. The product is easily understood, mounted, and used by naive users: the long treadles afford stepping, and the T-bar permits holding for balance. In case of damage or component failure, the pump is designed to be repaired by hand without the support of tools. The pumps are also portable, so they can be easily relocated in the field and taken in at night to prevent theft.

❸ Fisher argues that fighting poverty with charity is at best ineffectual, and at worst counterproductive. Charity, he argues, is invariably unfair, unsustainable, unappreciated, and breeds a culture of dependency. The name "MoneyMaker" exemplifies his belief that the best strategy—perhaps the only effective strategy—to defeat poverty is equipping entrepreneurs with the tools they need to make money. Consistent with this belief is the practice of selling MoneyMakers versus giving them away, for about $100. The average time for purchasers to recoup their investment is six months.

❹ The design is simple and durable, lacking screws, nuts, bolts, or fasteners of any kind. Parts are made of inexpensive locally available materials and created by indigenous fabricators. Relative to other treadle pumps, the treadles of the MoneyMaker are shorter and sit lower, modifications to reduce hip swaying for women users, which could be perceived by many cultures as being too provocative.

❺ The pump can be attached to either pipe or hose. Water is drawn in through one side on the upstroke, and expelled from the other side on the downstroke. The current model, Super MoneyMaker, can draw water up from 23 feet and has a total pumping head of 46 feet.

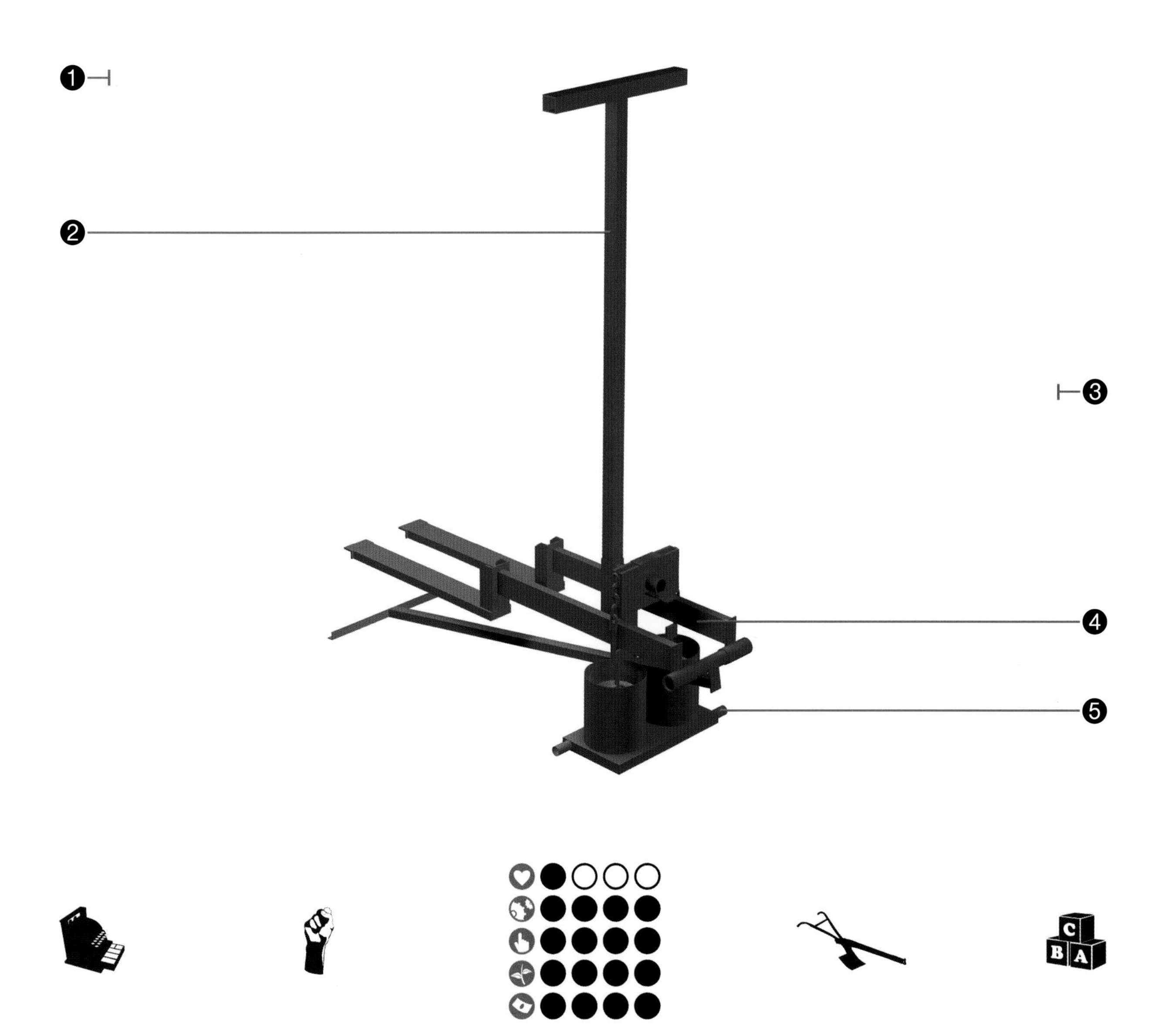
1
2
3
4
5
C
B
A

Mouse M0100

Dean Hovey, David Kelley, and Apple, 1984

❶ The mouse, the name inspired by the resemblance of early models to laboratory mice, had long been invented prior to its adoption by Apple in the early 1980s. But, as with the graphical user interface, it was Apple that transformed a largely experimental input device into a practical consumer-based product. In a textbook example of rapid prototyping, Apple commissioned pre-IDEO David Kelley and Dean Hovey to set about transforming a niche $400 Xerox PARC mouse into a mass-market $25 mouse—while also improving usability and reliability. Rapid prototyping is messy, based on developing numerous makeshift models versus formalized specifications. The premise is that each prototype teaches the designer something—problem, idea, insight—that likely could not have been learned without making and playing with a physical model. Accordingly, a series of rapidly developed prototypes can accelerate the learning curve in product design in a way no other process can. Dean Hovey comments on the proto-mouse: "The first place I went was to Walgreens, and I bought all the roll-on deodorants I could find on the shelves. They had these plastic balls in them that roll around. Then I went over to the housewares area and bought some butter dishes and plastic things that were about the size I might need to prototype something. Over the weekend I hacked together a simple spatial prototype of what this thing might be, with Teflon and a ball. The first mouse had a Ban Roll-On ball."

❷ The mouse introduced with the first Macintosh, the M0100, bears more than a modest resemblance to the Macintosh itself—same material, color, texture, and relative geometry. Edges bear a thick chamfer to improve hand fit and match the distinctive chamfer of Macintosh case. The result is continuity and coherence with the whole.

❸ Although earlier models of the mouse possessed more buttons, designers ultimately elected to favor ease of use over increased functionality and gave the Macintosh mouse one button—though the internal battle at Apple over the number of buttons purportedly lasted months. The debate is complicated because there is an intrinsic tradeoff between flexibility and usability, and the trade is mixed across multiple usability dimensions. For example, an input device with more controls can perform more actions directly, reducing the number of steps required to complete a given task. Direct actions and reduced steps are indicative of good usability. However, by virtue of the increased number of controls, perceived difficulty, learning curve, and error rates increase. High perceived difficulty, steep learning curve, and increased error rates are indicative of poor usability. The key variable in determining the optimal tradeoff across these dimensions is the degree to which prospective users are familiar with the way a product is to be used. For the overwhelming majority of Apple's target market, the mouse was a new kind of input device linked to a new kind of user interface. In this context, favoring simplicity over functionality was a good trade. As target users become more sophisticated, however, additional features and functionality become essential and can be added with little risk. If you are going to rely on one button, it needs to work well, and the button on the Mac mouse does not disappoint. The button pull is sufficiently heavy to resist inadvertent clicks, a small but essential feature that enables users to rest their finger on the button while positioning the cursor, yet sufficiently light to allow effortless activation when the intention is clear. Once pressed, visual, auditory, and tactile feedback is distinct.

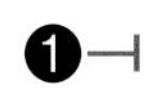

3
2

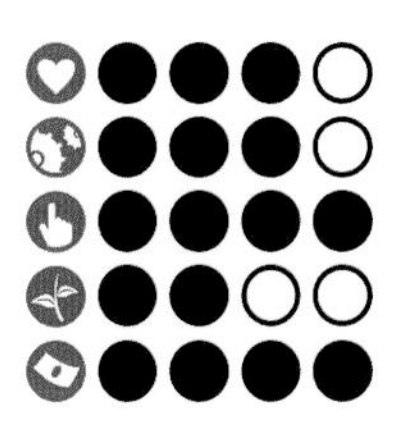

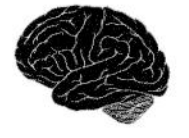

Museum Watch

Nathan George Horwitt, 1947

❶ A few thousand years ago, somebody put a stick in the ground and watched its shadow move with the apparent motion of the sun. This observation would ultimately lead to sundials, which enabled a gross measure of time. The number twelve held special significance to the peoples of that period, due to the fact there are twelve lunar cycles per year, and the number twelve became the standard number of divisions on sundials for the lighted day. The notion of sixty minutes in an hour and sixty seconds in a minute was borrowed from the base-sixty Sumerian number system. The sundial, however, would never be able to tell time at night. Timekeeping would need to become self-driven to overcome the darkness. Sundials would give way to water clocks, water clocks to mechanized clocks, and mechanized clocks to mechanized watches. The timepieces doubled up the twelve-hour cycle: twelve increments for day and twelve increments for night. Precision improved. A third hand was added to measure seconds. Lavish ornamentation soon adorned the hands and faces. Then, thousands of years from the time the first stick went into the ground, an industrial designer with a penchant for minimalism would halt this trend toward complexity, subtract out the superfluous, and once again pay homage to the sun. Nathan George Horwitt comments: "[I am a] functionalist designer with little understanding of decoration. To offset the clutter on the face of a watch, I carried the simplification of the numberless dial to three indices—a lunar dot at the top to mark the twelve, and the two hands."

❷ The Museum watch is the first watch to be inducted into the design collection of the Museum of Modern Art, which is the basis for its name. The watch has but one absolute reference point—a gold dot at the twelve o'clock position—which provides an orientation cue for the two hands. The design of the watch is perhaps as minimalist as it could be and still be considered a watch. The simplicity of form in conjunction with the gold on black elements is undeniably elegant, though, perhaps due to its marketing, is often considered pretentious—a fact strangely at odds with its sophisticated but humble aesthetic. The watch is often criticized for being difficult to read, and the criticism is valid when very precise times are required. It is important to remember, however, that most human activities revolve around the hour and half hour, and occasionally the quarter hour and three-quarter hour. For these hand positions, which are strongly horizontal and vertical in their orientations, the absence of numbers around the perimeter has little impact on the rate and accuracy of time telling. Humans visually process line orientations in the horizontal and vertical dimension very efficiently, thus subtracting elements from these axes is harmless. Additionally, the simplicity of the design offers something that more precise digital watches do not: context. The hands of the Museum watch give users a spatial reference of time, like a pie chart, which is more meaningfully understood than a point-in-time digital reference to the hundredths of a second. The Museum watch continues to be instantly recognizable, iconic, and timeless even by contemporary standards. Too bad future generations may fail to appreciate the genius of this design due to an inability to read analog time.

1
2
MOVADO

Natural Nurser Baby Bottle

Whipsaw for Adiri, 2008

❶ Baby bottles have come a long way since the leather-teated cow horn feeders of the middle ages, and the Natural Nurser Baby Bottle takes them a little further. With its anthropomorphic form and bisphenol-A- and phthalate-free construction, the bottle aspires to become a safe and viable sensory substitute for breast-feeding babies, while also improving upon the convenience and usability of traditional baby bottles. Dan Harden, president and chief designer of Whipsaw, comments: "A human breast is functional for breastfeeding and beautiful at the same time. Women who can't breastfeed, or want to combine breastfeeding with bottle-feeding, want the next best thing ... The design innovation is in the integration of soft and rigid plastic properties into one uniform part that offers a soft end for baby and a bottle that's easy to hold and fill for the parent."

❷ The form captures the organic aesthetic of a mother's breast, but with sufficient abstraction to look more like a sculptural rendering than a misplaced prosthetic. The bottle comes in three colors, each corresponding to different rates of flow: white for newborns, blue-purple for babies six to twelve months, and orange for babies one to two years. The colors are translucent and promote a soft, gentle affect for the bottle—it looks babyish, not bottlish. The tactile experience is soft and grippy, and though not the same as skin, pleasant. The nipple cap warms with the milk, and in terms of force feedback is very similar to a real breast. The bottle has few parts and correspondingly few seams, making it simple to clean. The price point is a premium, costing three to four times more than a conventional bottle—and additional multiples if you upgrade to the different models based on age.

❸ The nipple cap is designed to simulate the look and feel of a real nipple, the primary benefit of which is minimizing nipple confusion (also known as *nursing strike*). While experts debate the existence of this condition, it is generally the case that artificial nipples require less sucking energy than real nipples to produce milk. As babies will quickly learn to associate the size, shape, and feel of a nipple with the feeding technique that gets the milk, it is not implausible that a change in these oral and tactile cues could create confusion. Accordingly, the Natural Nurser should, at least in principle, facilitate the transition between breastfeeding and bottle-feeding. The logo and volume markers are blind debossed in the overmold, an elegant but potentially troublesome low-contrast marker system.

❹ The bottle is filled while upside down, its bottom cap removed and cover on. The process is simple, but although theoretically this configuration should not leak, it often does—the design could be greatly improved by making the nipple self-sealing and eliminating dependence on the cover. The bottom cap contains an airflow vent to prevent bubbles and air ingestion during feeding. When babies feed, the liquid in their bottle gets replaced by air, which in conventional bottles causes nipple collapse. As babies suck harder to get liquid through the collapsed nipple, they invariably swallow air. The Natural Nurser's vent lets air slowly into the bottle as babies drink, preventing the nipple from collapsing and minimizing the occurrence of gas and colic. However, the tradeoff is a bottle that dribbles all over the place when inverted—perhaps a small price given the benefit.

1
2
3
4
adiri

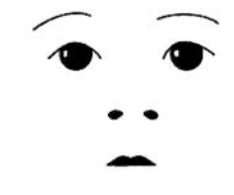

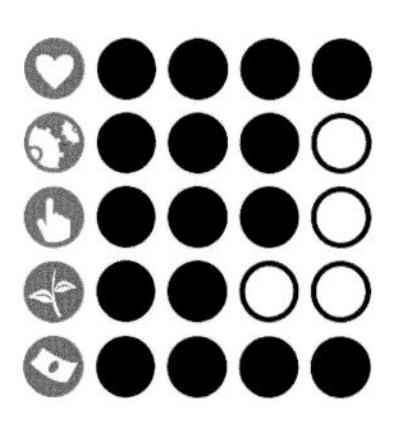

No. 5 Flacon

Coco Chanel and Ernest Beaux, 1921

❶ When Gabrielle "Coco" Chanel decided that she wanted to launch an eponymous perfume, she turned to noted chemist and perfumer Ernest Beaux to produce samples that were modern, original, and decidedly nonfloral. Beaux, drawing on an olfactory experience from a military expedition in the Arctic Circle, modeled the prototypical fragrances on the scent emitted by the Arctic lakes and rivers during the midnight sun. Beaux presented twenty-four samples to Chanel in numbered bottles for review. She chose the fragrance in bottle number five, which became the basis of the product's name. Coco Chanel's explanation: "I want to give women an artificial perfume. Yes, I really do mean artificial, like a dress, something that has been made. I don't want any rose or lily of the valley, I want a perfume that is a composition."

❷ The bottle merges the basic form and transparency of an apothecary bottle with the classic rectilinear lines of Greek architecture. It was inspired by the geometry of the Place Vendôme, a public square in Paris that matches the bottle's basic dimensions down to the chamfered corners and neck. The perfume itself was the first non-single-note floral scent, composed of over one hundred elements, including the first synthetically replicated molecules in a perfume, called aldehydes. The result is an abstract fragrance—a fragrance that is distinct, memorable, but malleable to the context. The label is printed using sans serif lettering on plain white paper, devoid of all ornament. The packaging is as spartan, resembling the generic black text on white boxes that one finds in duty-free shops. The plain packaging has certainly not impeded the products success: it is estimated that a bottle of Chanel No 5 is sold every fifty-five seconds somewhere in the world.

❸ The original flacon was more modern and minimal than the more iconic and recognizable form that we know today, possessing unfeatured edges and rounded corners, and a tiny lid matched by an equally tiny label. Its stark minimalism and right angles masculinized the bottle, mirroring the changing attitudes and fashions of women at the time. For nonfashionista readers, Chanel is known for creating women's attire that is simple, comfortable, and revealing. She once stated, "I gave [women] back their bodies—bodies that were drenched in sweat, due to fashion's finery, lace, corsets, underclothes, padding." She is popularly known for her minimalist suits and dresses (e.g., "little black dress") and women's trousers. The early success of her simple, masculine bottle with its complex scent was not, however, a function of women being naturally drawn to these kinds of forms and fragrances—quite the opposite, in fact, as women tend to prefer rounded forms and naturally derived scents—more likely it was what they represented: progress in civil rights and suffrage for women around the globe. Fashion trends of the period, particularly flapper fashion and culture, reflected this increased equality through more masculinized female preferences. The bottle would evolve soon after its introduction, however, no doubt a response to the overly austere design and the emerging Art Deco movement. Most of this evolution occurred in the first three years: thickening the glass, adding chamfers and bevels, increasing the size of the label, and transforming the lid from a nondescript stopper to an ornate diamond-cut crown. These refinements gave the flacon a more feminine, gemlike appearance, an aesthetic that would ultimately become the icon for the brand and make it the best-selling perfume in the world with estimated sales over $100 million annually.

1
2
3
Nº5
CHANEL
PARIS

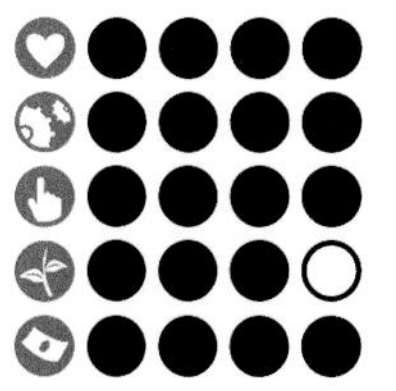

Officer's Knife

Karl Elsener, 1897

❶ The Swiss Army knife is legendary. Astronauts have used the knife to conduct makeshift repairs aboard the Space Shuttle. Doctors have used the knife to perform emergency amputations. Mountaineers have used the knife to dig their way out of avalanches. Skydivers have used the knife to cut themselves free of snarled parachutes. And no less than MacGyver himself (popular 1980s television character who used brain over brawn to fight crime and help the helpless) used the knife as his tool of choice to solve all manner of seemingly intractable problems. This success story, however, had an inauspicious beginning. In 1891, Karl Elsener and a number of fellow craftsmen founded the Swiss Cutlers Association to compete with the German knife industry for the Swiss Army knife contract. They produced the Soldier's Knife, a precision crafted multitool equipped with a blade, screwdriver, awl, and can opener. The knife focused on function—it was, after all for soldiers—but it was plain looking, heavy, and more expensive than the German offering. The Swiss army selected the German-crafted knives, and the members of the Cutlers Association went out of business one by one, including eventually Elsener. He fended off creditors and took out loans from family members to try again. He had observed that army officers were very conscious of their appearance, preferring clothing and equipment that differentiated them from the lower ranks. Targeting these more fashion-minded soldiers, Elsener introduced a more streamlined design with six tools and dubbed it the "Officer's and Sport Knife," commonly referred to as the "Officer's Knife." The repositioned product was a hit, and officers flocked to cutleries to replace their now comparatively staid army-issued pocketknives with these new, more elegant officer-class pocketknives.

❷ The Soldier's Knife used a dark wood for the handle. By contrast, the Officer's Knife used a lighter, redder wood, giving the knife a brighter and more expensive appearance than its predecessor. When the Officer's Knife was made available for sale to the general public, its handle was made pure red to contrast with snow. The red color would ultimately become the standard color for the handles of virtually all Swiss Army knives. The cross-on-shield logo was added in 1909 to emphasize its Swiss parentage.

❸ The knife was streamlined and tools added. The addition of the smaller blade and corkscrew tool is particularly significant, as it tapped into the elitist high-culture of officers of that period. The small blade was useful for scraping off mistakes in paperwork handwritten by pen, an administrative necessity for officers when processing paperwork. The corkscrew is obviously for celebrating and socializing over wine, again predominantly officer duty.

❹ The knife design suffers the same basic design challenges then as now: when folded in, the tools are hidden and difficult to discriminate; the tools are difficult to deploy, often requiring multiple tools to be pulled out to get the one in need; due to its relatively small size, smooth handle, and multiple sharp implements, it is not uncommon for the hand to slip with vigorous use—a serious flaw in emergency situations where injuries can have catastrophic consequences. Despite these deficiencies, the knife has always had a very strong function-portability ratio, making it an invaluable tool—until, that is, the terrorist attacks in the United States on September 11, 2001. That wildcard event hit the product where it hurt most: its portability.

1
2
3
4

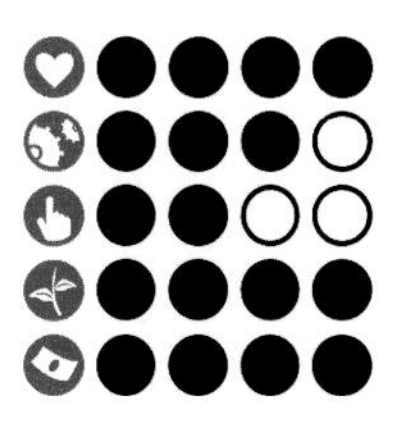

O-Series Scissors

Olof Bäckström for Fiskars, 1967

❶ The use of scissors predates written history. The design has evolved from "spring scissors," comprised of opposing blades connected by a curved metal piece, to the now more common "pivot" variety in which the opposing blades pivot around a fixed axis. After that, however, not much has changed—which means that the fundamental design of scissors, as we know them, is the product of about two thousand years of continuous refinement. Olavi Lindén, chief designer at Fiskars, comments: "All tools exist already: the basic implements, the axe and the knife and the spade. The same tools are still made at Fiskars as in primitive cultures. But they cannot be replicas of something old, they have to be in some way modern and in some way timeless, to live as long as possible. Our products have a 500,000-year life cycle."

❷ There is little difference between the current model of Fiskars scissors and the 1967 O-series, a testament to both the design of the Model O and the legacy of design that preceded it. The bright orange handles are effective at popping out of the invariable clutter of the cutting room, facilitating their location and give the scissors a fun, lively affect. The blades are manufactured to exacting specifications, a level of precision that can be verified by ear when the scissors are opened and closed in the air: an accelerating "swick" sound as the cutting point moves along the blade all the way to the tip. The 1967 version was limited to right-handed users only. A version for left-handed people was made in red in the 1970s. Loyalty to the brand is legendary, which is understandable given that scissors are the principal tool of the trade for tailors and seamstresses. And woe to the child who cuts materials other than fabric with a parent's Fiskars.

❸ Contrary to popular belief, the basic ergonomic form of the O-series handles was not new to this model. The curvy handles with their finger-friendly holes had actually been developed in 1880 for professional-grade tailor's scissors, originally cast in brass and joined to wrought-iron blades. The scissors were without parallel in terms of their performance and comfort, but their high cost of production restricted distribution to professional tailors and seamstresses. That is, until the early 1960s, when Olof Bäckström was commissioned to explore creating a low-cost plastic version for the mass market. After a number of years of development, Bäckström sent orders for the first plastic prototypes to be produced in black, red, and green. The plastic works was finishing a production run of Fiskars popular plastic orange juicer, and residual material had been left in the machine. Rather than let it go to waste, the frugal operator decided to use the remaining orange plastic to create prototypes in an unrequested fourth color. The prototypes were delivered, and the scissors with the black and orange handles were deemed the favorites. Fiskars then decided to take an internal vote to make a final decision between black and orange, and the prototype with the orange handles prevailed. One can only wonder if the scissors produced in 1967 would have achieved their current level of success and iconic status had it not been for the serendipitous event and subsequent selection process that so differentiated the aesthetic of the handles. It seems unlikely. Lessons: 1. great design is often as much a function of luck as expert work—that is, for those willing to treat the unexpected as opportunities versus threats; and 2. there is no test for aesthetic preference like an empirical test.

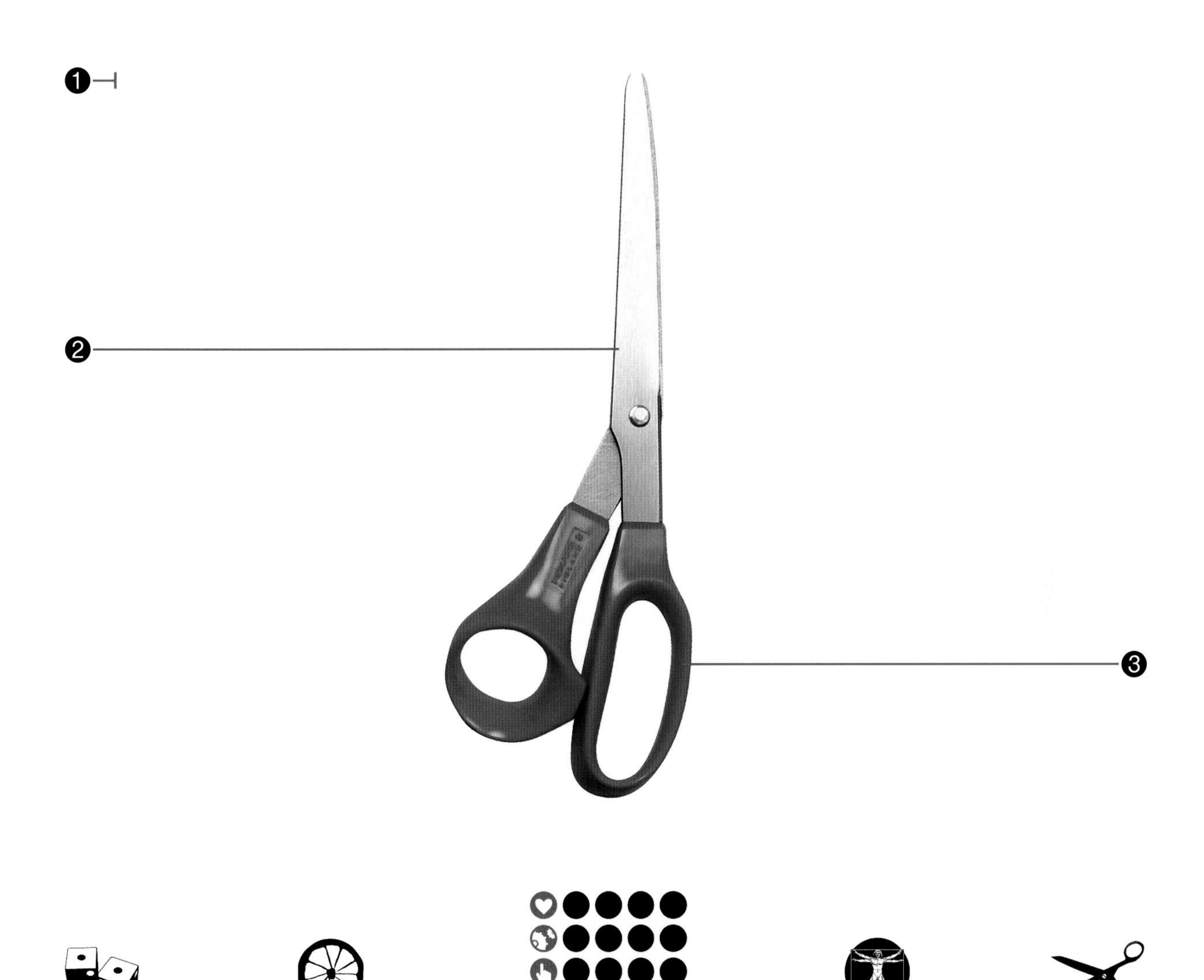

PalmPilot

Jeff Hawkins and Palo Alto Design Group, 1996

1. Extraordinary products are invariably born of extraordinary generalists supported by extraordinary specialists. In the case of the PalmPilot, the extraordinary generalist was none other than Jeff Hawkins, entrepreneur, inventor, and self-made neuroscientist with design and fabrication skills honed from early childhood. Jeff Hawkins comments: "[My family] designed boats, built boats, lived on boats; most of my family lives on boats today. My father was sort of this consummate inventor. If you've ever seen the movie or the play *You Can't Take It With You* with Jason Robards, it's about this crazy house where they had a great time, but it's really crazy... I tell that story because all those experiences, you know, learning to use shop tools, learning to use fiberglass and screws and stuff, is actually very useful in building little computers. Not many people in my business can get involved in as many aspects of product design as I do."

2. The four-shade monochrome display is basic but functional—a small price considering the battery life is about one month with standard use. Resolution is 160x160 pixels, which is paltry by contemporary standards but more than adequate for the productivity focus of the device.

3. The form factor is reverse engineered from Hawkins' shirt pocket. The first prototype—a whittled down piece of wood using a chopstick for a stylus and a printout of an interface taped to the front—was carried around by Hawkins for weeks, during which time he would frequently whip it out and interact with it. The product's utility, efficiency, and ease of use speak to the understanding of the requirements achieved during this primordial period of concept prototyping. Its effect, however, is akin to that of a tool, a wrench—a utilitarian device with no transcendent aesthetic.

4. The PalmPilot used a stylus for tap-inputs and text entry. Though small by traditional pen standards, the stylus performs well and has proven itself an intuitive input device. When not in use, the stylus was stored in a spring-loaded silo molded into the case.

5. The operating system was built around handwriting recognition software named "Graffiti," which required users to learn a kind of shorthand, and then write this shorthand letter by letter into the "Graffiti area." In the wake of similar devices that failed due to poor handwriting recognition, this was a tough sell, and most gave the PalmPilot dim prospects for success. But succeed it did, shipping over a million units in eighteen months and attaining a seventy percent market share. So how is it that the PalmPilot succeeded when devices with more sophisticated handwriting recognition, such as the Apple Newton, failed so profoundly? Size and price definitely played a role—the PalmPilot was less than half the price and size of the Newton—but there is also a lesson here about expectations management. The PalmPilot set and exceeded modest expectations. The Newton set and failed to meet high expectations. It is always better to set modest user expectations and overdeliver, than to set high user expectations and come up short.

6. The four physical application buttons enable quick and direct access to key productivity applications, which can be reassigned by the user. Additional applications are represented iconically in the display and are accessible using the stylus.

PalmPilot
Memo 5 of 6
Unfiled
Notes from guitar lesson on 10/14.
Make sure that you tune your guitar before you try to play the instrument.
Pay attention to hand position. For classical guitar, the left thumb should be centered on the neck of the guitar.
Done
Details
APPLICATIONS
CALCULATOR
MENU
FIND
abc
123
1
2
3
4
5
6

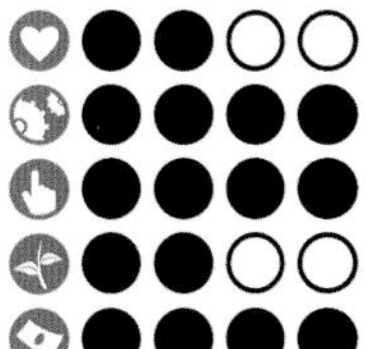

Pencil Sharpener

Raymond Loewy, 1933

❶ Considered the father of industrial design, Raymond Lowey redesigned everything from Coke bottles to bullet trains. One key to his success was a surprisingly formalized method for balancing the progressiveness of his designs against cultural norms, maximizing prospects for acceptance and commercial success. Raymond Loewy comments: "There seems to be for each individual product (or service, or store, or package, etc.) a critical area at which the consumer's desire for novelty reaches what I might call the 'shock-zone.' At that point the urge to buy reaches a plateau, and sometimes evolves into a resistance to buying. It is a sort of tug of war between attraction to the new and fear of the unfamiliar... When resistance to the unfamiliar reaches the threshold of a shock-zone and resistance to buying sets in, the design in question has reached its MAYA—most advanced yet acceptable—stage. We might say that a product has reached the MAYA stage when thirty percent (to pick an arbitrary figure) or more of the consumers express a negative reaction to acceptance. If design seems too radical to the consumer, he resists it whether the design is a masterpiece or not. In other words, the intrinsic value of the design cannot overcome the resistance to its radicality at the MAYA stage."

❷ The pencil shapener's form is streamlined, futuristic, perhaps modeled after a dirigible, which had become an icon of technological progress until the Hindenburg disaster in 1937. Once the product is identified as a pencil sharpener, its use is intuitive: insert the pencil into the hole, turn the crank. Pencil shavings are tidily collected in the base, which is removable for cleaning.

❸ The design never went into production, and its function never refined to work properly—but function has nothing to do with this design. His patent filing confirms this, without apology: "Be it known that I, Raymond Loewy, a citizen of the Republic of France, residing at New York, the county of New York, in the State of New York, have invented certain new, original, and ornamental Design for a Pencil Sharpener...." The design of the pencil sharpener is more about what is says than what it does, as what it does is mundane, so the product is used to tell a story rather than to sharpen pencils. Function in this context is essentially a commodity, a secondary consideration in the overall design. Accordingly, the product seeks not to merely sharpen pencils, which it needs only do with some minimal facility, but to transform an everyday object into something people want to proudly display on their desk versus hide away in their desk drawer. Function here is used to help users rationalize the purchase; to help justify a visible placement in an office without appearing ostentatious, as it might with a piece of modern art; to pronounce a level of taste and sophistication without appearing boastful. But function is not the reason consumers buy a pencil sharpener like this. One does not have to squint to see the quasifunctional, streamlined DNA of this product in contemporary products like Starck's Juicy Salif. Starck asserts that the primary design objective of the Salif is to begin conversations. One suspects that Loewy, pencil sharpener proudly placed at a strategic location atop his desk, would share the sentiment.

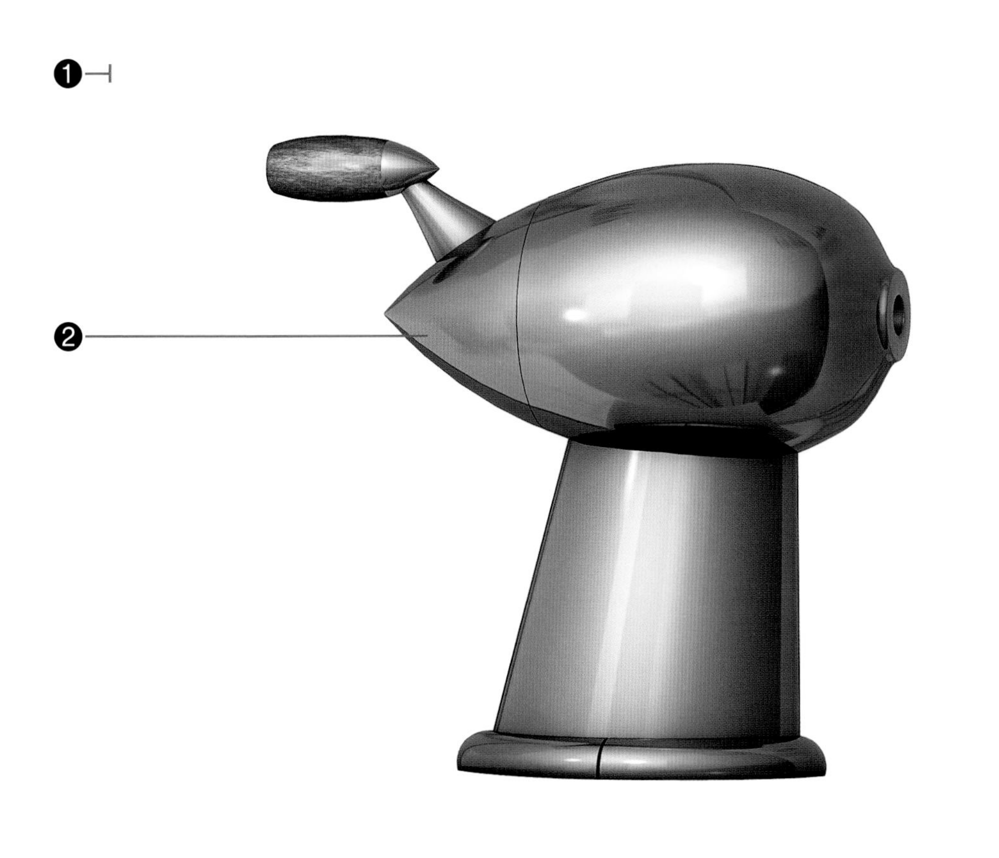
1
2

3

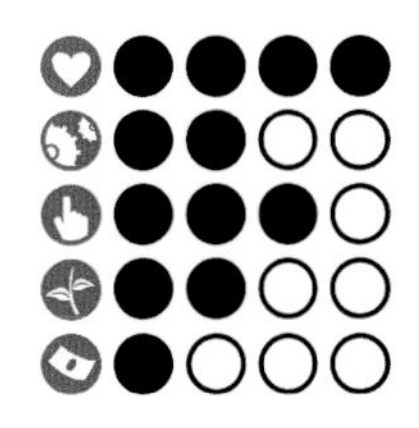

100

Phonosuper SK 4

Hans Gugelot and Dieter Rams for Braun, 1956

❶ After hitting what one could call a "design block" in creating a new generation of low-cost radiograms, Braun designers Fritz Eichler and Dieter Rams called on Hans Gugelot for assistance. Gugelot, a teacher at the Ulm School of Design (HfG) in Ulm, Germany, brought his functionalist aesthetic and mastery of manufacture to bear on the problem, a collaboration that ultimately produced the SK 4, and a philosophy Rams would articulate in his "Ten Commandments of Design" some years later. Dieter Rams's First Commandment is particularly apropos: "Design is innovative. It does not copy existing product forms, nor does it produce any kind of novelty just for the sake of it. The essence of innovation must clearly be seen in all of a product's functions."

❷ This radio-record player combination looks organized, technical, like a machine, a stark contrast to the furniture-like style of appliances of the day. By modern aesthetic standards, however, the device appears sophomoric, like a misplaced prop from *2001: A Space Odyssey*—an example of what people in the mid-twentieth-century thought the future was going to look like. Controls are located on the top of the unit versus the front, optimizing their visibility and access. White space is ample. The interior body is a single piece of sheet steel bent to form, cutouts for speakers and cooling, sandwiched between two wooden panels—a configuration that simplified manufacture and minimized cost. Sound quality of the player was good, but not on par with its high-end competitors. The product's most compelling element, its transparent lid, was not born of prescience or planning, but of the need to address vibration problems caused by the original sheet steel cover.

❸ The transparent acrylic lid was an innovation in its day, giving the unit a novel appearance and implying that it was an object to be featured, showcased, not hidden away like other players. As it was something new, however, consumers needed a way to understand and talk about it, and the analogy that seemed to stick was "Snow White's coffin," after the glass coffin described in the Grimm Brothers' fairy tale. Although the origin of this nickname is unknown—Rams has speculated that it may have come from a competitor—the player's technical-sounding name did little to discourage the public's search for a vernacular alternative.

❹ The top of the unit has a phonograph, grouped by a gray rectangular surround, and a set of selection and radio controls, grouped together into a rectangle by their relative proximity and alignments. Correspondingly, the front of the unit has two rectangular sets of horizontal grilles, which are positioned to visually relate to the phonographic and control groupings. The phonographic grille set is left aligned with the phonographic grouping, but the grilles do not extend sufficiently to right align with the right edge of the gray surround. Continuity and integrity are broken. The same problem exists with the control grouping-grille alignment on later models. In early models, the control grouping was right- and left aligned with the control grilles. However, as controls were added and the grouping widened, the grilles were not updated, making the grille set fall just short of left alignment. No doubt the grilles are as wide as functionally required and no wider, but sacrificing this very visible and powerful alignment cue for the sake of functional purity illustrates one of the reasons there is now a *postmodern* age.

2
3
4
BRAUN

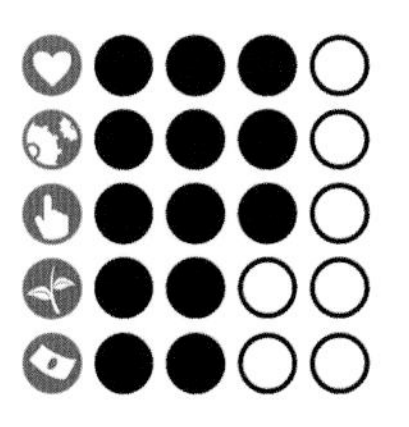

Pleo

Caleb Chung and Ugobe, 2007

❶ An animatronic dinosaur from the creator of Furby, Pleo is an example of a new class of entertainment products referred to as "social robots." Social robots communicate and interact with humans in behaviorally plausible ways. Although Pleo has a ways to go before approximating a real cat or dog, it is not as far as many think—when Pleo tilts its head and "purrs" in response to a head scratch, it is easy to forget that it is an artificial being executing preprogrammed social behaviors. Caleb Chung comments: "All that matters is what the user perceives. Once you create organic movement and add just enough emotive cues, people will suspend disbelief—they'll fill in the rest and fall in love with Pleo."

❷ The robot is modeled to resemble a baby *Camarasaurus*, a genus of quadrupedal, herbivorous dinosaurs common to North America in the Late Jurassic period. The choice is interesting, leading one to wonder for whom this robot is designed. As an herbivore, this particular species lacks the features to trigger predator fixation in male children—that is, it lacks angular features, forward-looking eyes, sharp teeth, and claws. Additionally, the creature is relatively featureless, making it uninteresting. If for whatever reason the species had to be nonpredatory, a more interesting form factor would have been a baby *Triceratops* or *Stegosaurus*. The reptile-like appearance and rubbery skin undermine nurture fixation in female children. Although the baby-face features are appealing, the cold, rubbery skin, rigid interior, and inflexible body make petting and cuddling an innately dissatisfying experience. The high price point limits access to affluent, geeky adults. The short battery life restricts its viability as a toy. For whom is this robot designed? Perhaps its creator, and disenfranchised AIBO owners.

❸ The legs, tail, neck, and head coordinate to create realistic movements, albeit disappointingly slow and jerky. Compared to the highly organic and purely mechanical movements achieved by Theo Jansen, for example, Pleo's moves are a poor approximation of what's possible. The whirring sounds of its many motors are audible, undermining the suspension of disbelief and signaling the amount of work being expended to achieve even rudimentary movements; a cost made further evident by its short battery life. Still, there is much promise in the motion technology. For example, Pleo can learn to optimize its walking motion for different surfaces: a Pleo that grows up on a shag carpet will learn to walk differently, stepping higher to overcome the pile, than a Pleo who grows up on hardwood floors, tending to drag its feet. Better than its predecessors, but still very first generation.

❹ The learning capacity and behavioral repertoire of Pleo is its most interesting and captivating attribute. Simulating three stages of development—hatchling, infant, and juvenile—Pleo grows cognitively to develop a unique personality based on its lifetime of social and environmental interactions. It has the capacity for a variety of emotional states, including happy, scared, surprised, and sad, to name a few, each yielding plausible behavioral responses—for instance, a happy Pleo coos, an agitated Pleo growls. The one behavior Pleo does not yet have is the capacity to self-charge, a capacity sorely needed for it to become a viable robotic pet. Perhaps Ugobe should offer an aquarium-like Pleopen, complete with flora and toys, which keeps Pleo active and charged. This would support a more cost-justifiable product story of Pleo-as-pet versus Pleo-as-gadget, making the artificial critter an exotic, entertaining, and maintenance-free alternative to fish, hamsters, and birds.

1
2
3
4

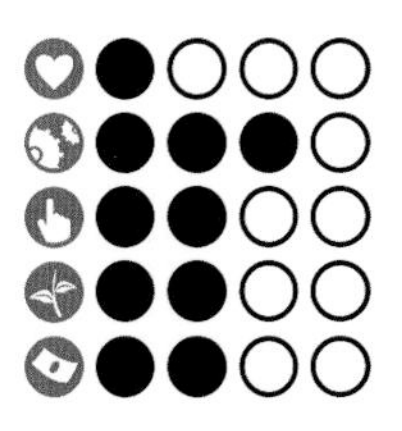

Pocket Survival Tool

Tim Leatherman, 1983

❶ The Swiss Army knife is unquestionably a classic of design and the seminal multitool, but in modern use one is quickly reminded that the knife was originally designed for soldiers in the late 1800s. Still useful, but perhaps not ideally suited for practical civilian use—a limitation that would ultimately be the inspiration for the modern Leatherman multitool. Tim Leatherman comments: "My wife and I decided to travel abroad in 1975. We were young—it was one of those budget trips, and we bought an old Fiat in Amsterdam for $300. I was carrying a Boy Scout–type knife and used it for everything, from slicing bread to making adjustments to the car. But I kept wishing I had a pair of pliers! During the trip—it lasted almost nine months—I had a piece of paper in my pocket where I listed ideas for new products, things I might work on back in the U.S. It was in a hotel room in Tehran that I started sketching a pocketknife that contained pliers. Once we got back to Portland, I asked my wife if I could build it—just one for me. I told her it would only take a month, and she got a job to support us... My month turned into three years. I learned that I'm not a very good inventor. I don't have much foresight. You know Marconi, who built some of the first radios? I've heard that before he picked up a pencil, he had the entire model envisioned in his mind. I'm not that way. It took a few months just to visualize each part of the knife."

❷ The Leatherman PST is constructed of high-grade stainless steel. Folding motions are firm but smooth. The device looks and feels dependable, indestructible—the kind of tool you want in an emergency situation. The moniker "LEATHERMAN TOOL" is machine etched into the handle, a dull and superfluous branding given the refined embossment around the primary axis. Basic operation is easily discoverable through the effective use of visibility and constraints.

❸ In the folded configuration, the PST looks like a metal folding yardstick. It is small enough for the pocket, but too large and heavy to be casually carried—it is more appropriately sized for the glovebox of a car, a backpack, or hanging from a lanyard. In the opened configuration, the PST becomes a pair of needle-nose pliers. This is the base functionality of the multitool, and as pliers go, its performance is equivalent to its unifunctional cousin. The base functionality of a multitool is of critical importance, as it defines the primary functionality of the tool and determines the real estate and design affordances of its secondary functionalities. The post-industrial world is replete with small cracks and crevices filled with wires, tiny fasteners, and mechanisms of various sorts. As a base function, pliers make sense, and offer a greater range of practical applications than a knife. And, unlike the Swiss Army variety of knives, the Leatherman PST is able to leverage the real estate provided by two handles, essentially doubling the physical envelope of a knife for additional tools. The PST also leverages the handles to improve safety. When a particular tool is deployed (e.g., a blade) the handles are designed to close, preventing the tool from closing and causing injury while in use.

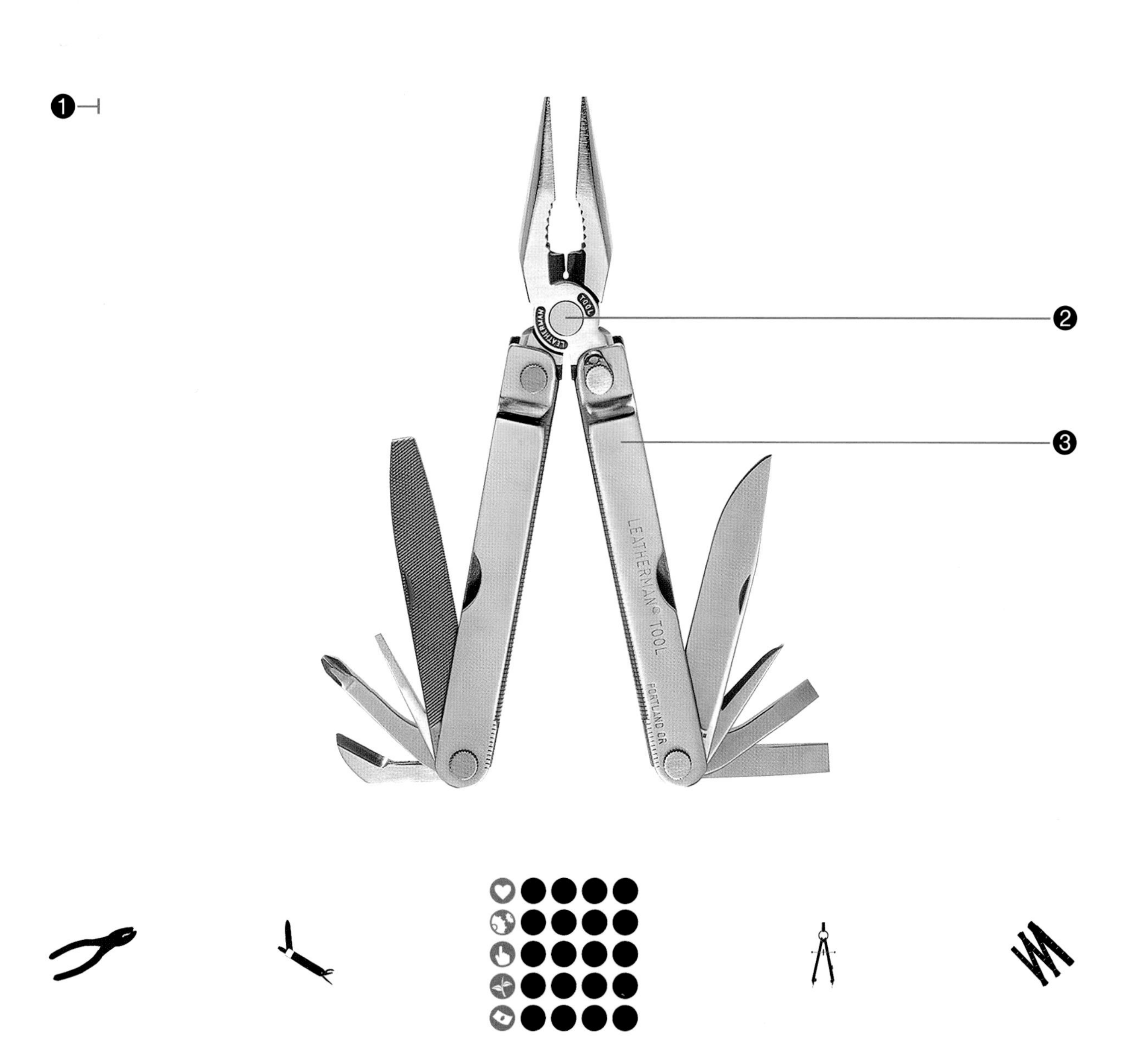
1
2
3
LEATHERMAN TOOL
LEATHERMAN® TOOL
PORTLAND OR

POM Wonderful Bottle

POM Wonderful, 2002

1. POM Wonderful, the name based on the "Wonderful" variety of pomegranate fruit, has single-handedly created the market for pomegranate juice in the United States. And the POM Wonderful juice bottle has played a significant role in establishing the brand as the premium provider in that market. Launched primarily with celebrity product placements, the brand has maintained its growth momentum with a simple advertising campaign reminiscent of the Absolut Vodka series, which prominently feature the bottle in a variety of playful contexts that highlight its beneficial qualities. The bottle at the time of its launch was unique in size and shape, and this differentiation was rewarded with sales. Fiona Posell, director of corporate communications for POM Wonderful, comments: "The bottle design was conceived and designed internally. The bottle represents our business—pomegranates—and the shape of the container was a crucial element to the design. It continues to leverage our brand differentiation goals."

2. The form is abstract and iconic, comprised of two stacked pomegranates. Highly differentiated relative to other juice bottles. The roundness of the form conveys a friendly affect, and the proportions are both snowman-like and feminine in nature—the waist-to-hip ratio of the bottle approximates 0.7, the aesthetically ideal proportion for a reproductively healthy female, and the chest-to-hip proportions are perfectly symmetrical, also a signature of health (e.g., assuming a waist of 24," the scaled measurements would be 34"-24"-34"). Originally available in glass, the company has since converted to plastic, a move to make the bottle lighter and more portable. The bottle affords secure gripping at both the neck and waist.

3. The head of the bottle is shaped to appear like the distinctive crown of a pomegranate, but in this purely abstracted form it looks like a monarchical crown, reinforcing the premium nature of the product. The bottle cap colors vary across the different flavors.

4. Type is screenprinted directly onto the transparent bottle, avoiding the concealment of paper labels and reinforcing the purity of the product—the bottle is transparent in the sense that it is see-through and that it has nothing to hide, conveying a sense of brand integrity. The red and white type contrasts strongly against the dark purple juice, and the branding is rich with meaning. The first term in the name, *POM*, is an abbreviation of the unwieldy *pomegranate*, a simplification that is also semihomonymous with the word *palm*—enabling the initial term to suggest both juice type and that the object is something to be held. The second term is a double entendre, describing the variety of fruit and conveying that the juice is something to be celebrated and savored. The primary typeface appears custom, echoes of Vag Rounded, with a heart substituting for the *O*. It, too, is friendly and playful. The heart, the most prominent element, emphasizes the core health claim—that is, pomegranate juice has heart-healthy benefits—and conveys a "love" affect. The last line of the messaging explicitly states the purity of the product at 100 percent, a key differentiator, distinguishing it from competitors who use less expensive filler juices. Although the labeling is minimalist, a tremendous amount of information is conveyed—every element in the label and bottle design conveys multiple synergistic propositions about the product, explicit and implicit, descriptive and affective, conscious and subconscious.

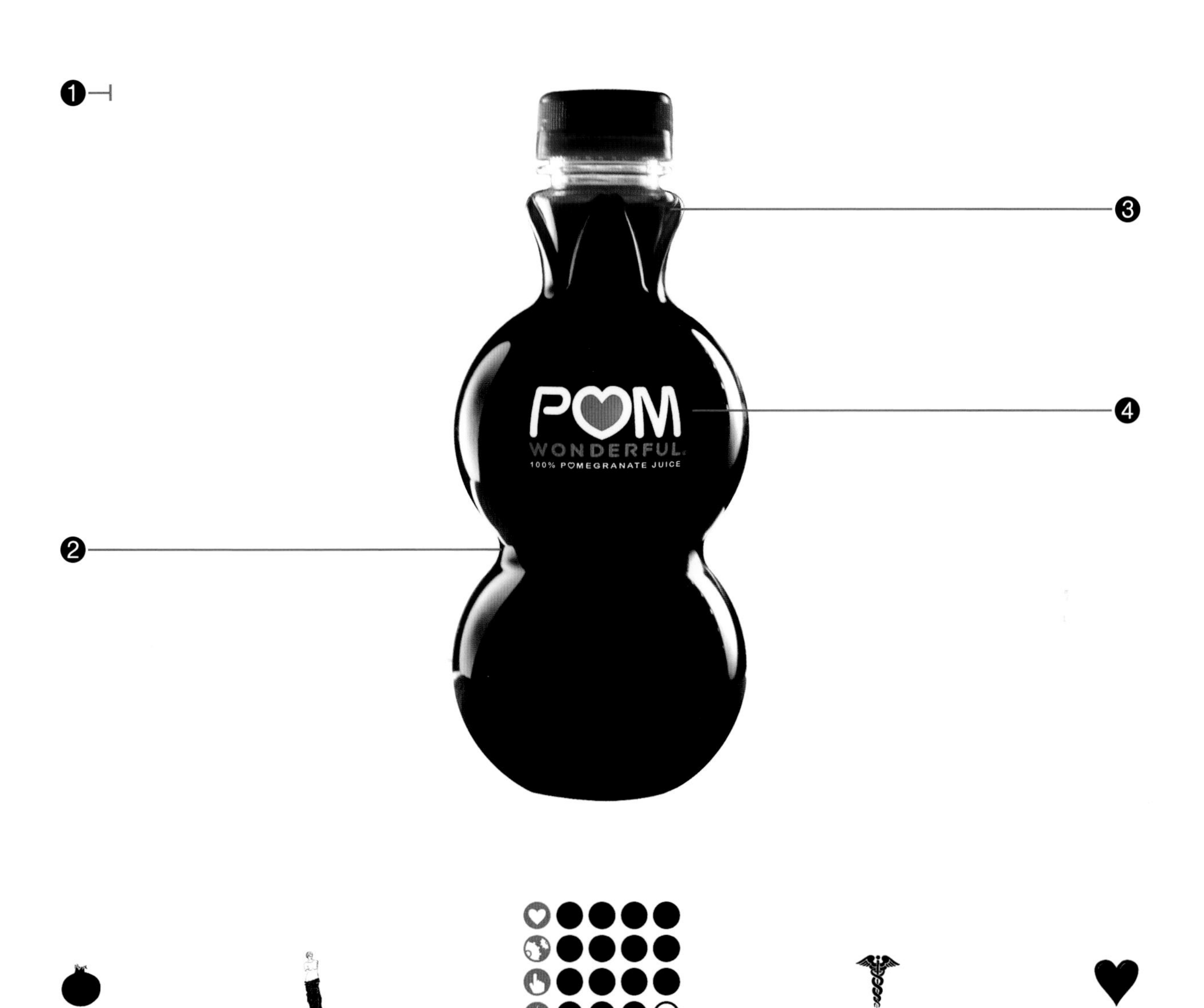
POM
WONDERFUL
100% POMEGRANATE JUICE

Post-It Note

Art Fry and Spencer Silver for 3M, 1980

❶ The Post-It Note began as the proverbial solution in search of a problem. Spencer Silver, a senior chemist at 3M, had developed a reusable, low-tack adhesive, but a marketable form of the product proved elusive. Art Fry learned of the adhesive at an internal 3M seminar, and realized it would solve a problem he experienced every Sunday: when he stood and opened his hymnal to sing at church, the bookmarks in his hymnal would fall out. The Post-It Note was born. Focus groups were conducted without samples, and consumer reaction was tepid. However, when follow-up studies were conducted with samples, consumers couldn't get enough of them, underscoring the divide that so often exists between user reports and user behaviors. Art Fry comments: "The Post-It was a product that met an unperceived need... If you had asked somebody what they needed, they might have said a better paper clip. But give them a Post-it Note, and they immediately know what to do with it and see its value."

❷ The original forms were rectangular, 3"x5" and 1.5"x2," and came only in canary yellow. Contemporary Post-It Notes come in a variety of shapes and colors, all of which are designed to fit in a pocket for portability. The small sizes are also an effective means of chunking thoughts and communications, facilitating both writing and reading notes. Simple shapes and color variety add order and structure to handwritten notes and are conducive to brainstorming and ideation—a value not immediately appreciated until contrasted with scrap paper, its notepad predecessor.

❸ The adhesive strip spans the width of the note on one side, giving it a clear top position. The adhesive itself was originally considered a failure, as the value of an adhesive traditionally corresponds to its bonding strength. It is a credit to Silver that he realized the potential value of an underperforming glue. Much of the appeal is the ease with which a note can be added, removed, relocated, and replaced—in the physical world, the Post-It Note is as close to an "Undo" function as you can get. In fact, one of its more popular uses is for editing large documents, as notes and commentary can be easily added and removed as revisions are made. This particular application has achieved infamy within government circles, where references to being FLYNned—an acronym for "fucking little yellow notes"—is often heard when note-ridden documents are returned to their author, with each Post-it Note representing a proposed change. The notes form a pad by binding to one another; parsimonious packaging which also introduces how the notes are to be used.

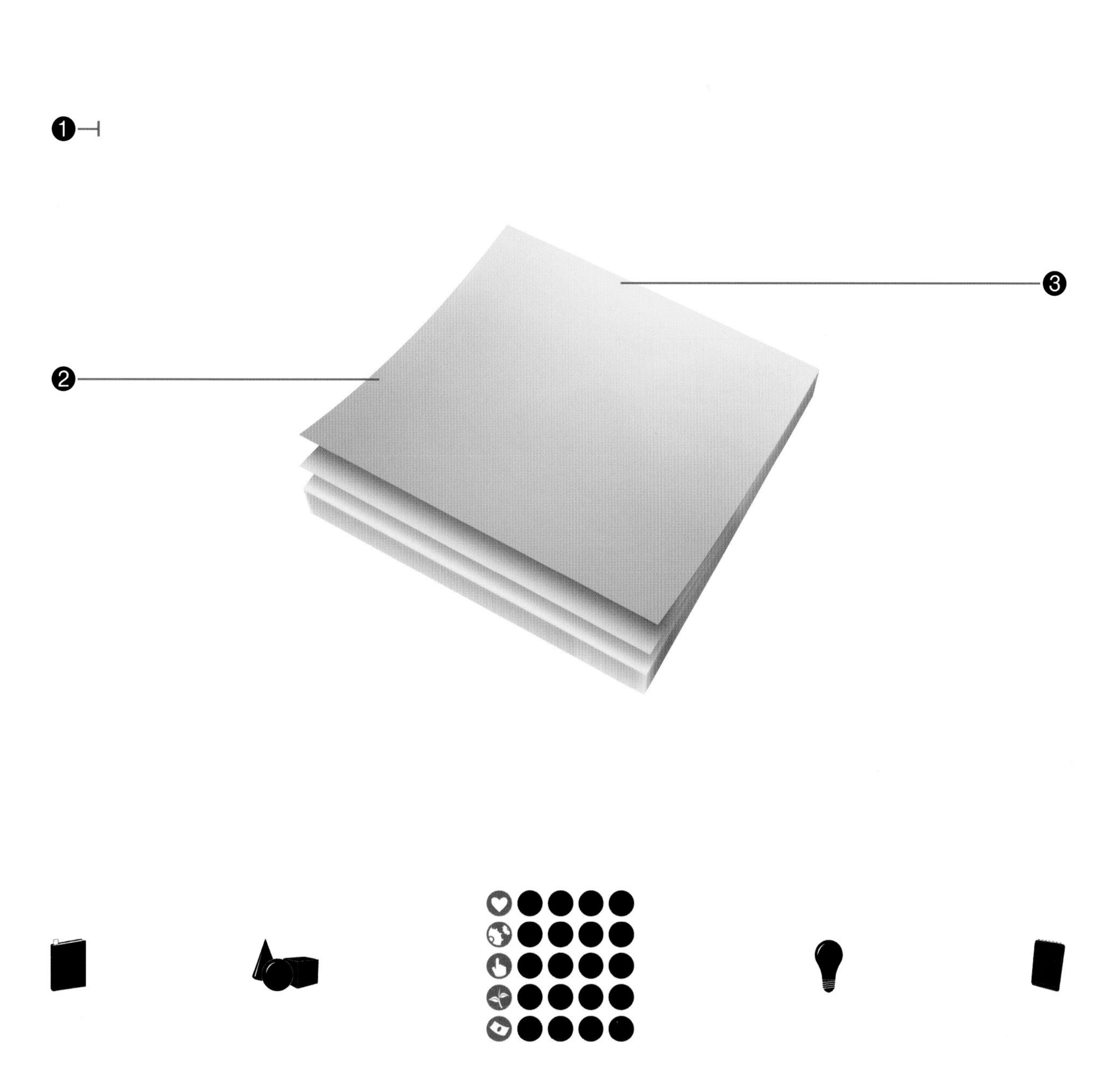

Pot-in-Pot Cooler

Mohammed Abba, 1995

❶ Successful design requires a deep understanding of the contexts in which a product will be used, and there is no substitute for getting in the field to observe and learn first-hand. It was this type of experience that enabled Mohammed Abba to identify a low-cost entry point for disrupting a very complex poverty cycle. Mohammed Abba comments: "I was at one time working as a regional consultant for the United Nations Development Program... I saw the hardships suffered by farmers. Farmers would produce a crop, but because this crop could not be preserved, it would spoil quickly, and as a result the farmers would throw out the crops. Or, if they did not throw them away, they would sell them below their production cost to middlemen who were coming from the cities who exploit their situation: 'You produce something, you have no way to preserve it, if you like, you can sell it to me for less than the cost of production. If you don't, you leave it and the crop spoils altogether.' So this is a very painful situation. This is what we call the virtual cycle of poverty, where you produce, you have no means of selling it, and the crop spoils. This discourages farmers from the practice of farming. As a result, so many able-bodied farmers migrate to the cities. In the cities, they are not skilled... so they end up going into a lot of other social vices in order to survive. This is where I decided to come in and help these people get a medium for preservation for their crops. In the process, I came up with the 'Desert Fridge' to help them prolong the storage life of their crops."

❷ The cooler is comprised of two unglazed earthenware pots, one small and one large, such that the smaller sits inside the larger. The space between the pots is filled with sand, and then the sand is saturated with water. Food is placed into the interior pot, and the top is covered with a wet cloth. Water slowly wicks into the pottery from the sand and evaporates when exposed at the inner and outer surfaces. This is where the cooling cycle kicks in. The outer air is hotter and drier than the inner air, and so moisture on the outer surface evaporates faster than moisture on the inner surface. This draws moisture (and with it, heat) outward, making the interior volume cooler. The insulating effects of the pottery, sand, and wet cloth cover help maintain the reduced interior temperature. The results are impressive, with temperature reductions inside on the order of 10–20°C below ambient. At this reduced temperature, tomatoes can last several weeks versus a few days. African spinach that wilts in just a few hours after harvest is preserved for up to two weeks. The ability to preserve the food in this way alleviates the pressure to sell crops immediately, which then enables farmers to better negotiate pricing, freeing family members—especially children—to engage in activities other than taking crops to market, like going to school. As men learn that they can make money farming, they are motivated to return home from the cities. And since the technology is based on a craft that is well accepted and supported by local potters, most of whom are women, women are empowered to sell food from their homes and break longstanding dependencies on men as the sole providers.

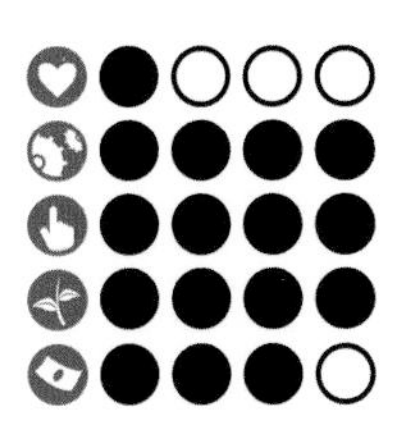

C
B
A

Power Mac G4 Cube

Apple, 2000

❶ A minimalist descendent of the NeXTcube computer, the G4 Cube followed the colorful line of iMac computers in heralding the return of Steve Jobs and good design sensibilities to Apple. The computer was not only unique in terms of form and material, it was very small relative to its computing counterparts and featured a number of engineering and usability innovations. Steve Jobs comments: "This is a stunning product. Quite possibly the most beautiful product we have ever designed. The computer is in an eight-inch cube. It's suspended in a stunning crystal clear enclosure ... Amazing cooling system through a center channel that cools the whole thing with air so that it runs in virtual silence and ... allows you to get at every component in just seconds."

❷ Per its namesake, the computer is cubical in shape, though if you consider the outer case, it is rectangular in profile. To designers, the Cube looks like a work of modern art. To nondesigners, it looks like a toaster, complete with a top-mounted DVD player to complete the effect. The toaster-like appearance is a problem, undermining the image of the Cube as a serious, high-end computer. Additionally, highly symmetric forms such as cubes tend to be perceived as aesthetically pleasing but generally uninteresting—to everyone but modernists, that is. In one study, for example, researchers gave users blocks of foam in various shapes and surveyed their preferences. The result was that users favored "dramatic" shapes in which at least one dimension was very different from the others. This suggests that in terms of mass-market form factor preferences, asymmetry typically trumps symmetry, and may explain, in concert with its toaster-like persona, the failure of the computer.

❸ The On/Off button is the only visible control of the computer—a small painted icon on the top surface of the case. The button is activated by touch, a prelude to the touch interfaces of future iPods and iPhones. Once it's touched, the only feedback is a glowing light that appears deep beneath the surface. Normally, a lack of tactile feedback is a deficiency, but in this case the effect is almost magical.

❹ The electronics are configured around a large, centralized heat sync, which conducts the heat toward the center of the cube. A carlike grille at the top of the Cube allows the heat to exit, creating a chimney effect, which sucks cool air through the intakes on the bottom. This convective process is sufficient to cool the computer, avoiding the need for noisy fans—an engineering achievement, but potentially problematic for users accustomed to stacking their papers on the nearest available flat surface.

❺ The clear case looks sophisticated and expensive. It is thick, solid, and aesthetically pleasing, but its rigidity results in the transmission of a surprising amount of DVD-drive vibration to the desk. Communication ports are concealed on the underside of the suspended cube, keeping the visible surface of the enclosure uninterrupted. The primary issue with the acrylic enclosure regards the formation of nonstructural cracks in its surface. Whether caused by manufacturing defects or poor material durability, the cracks hit the product where it hurts the most: its aesthetics. The problem was never rectified prior to the discontinuation of the line, and no doubt contributed to its demise. Act 3 for the cube form: the Apple Store in New York City, which shares more than a passing resemblance to the crystalline G4 computer.

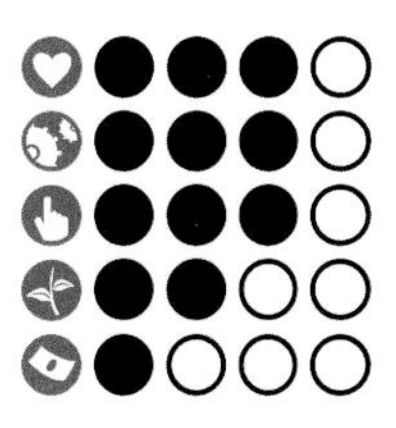

Q-Drum

Hans Hendrikse and Pieter Hendrikse, 1993

❶ For those living in developed nations, it is easy to forget that a significant percentage of the world population still engages in a daily struggle to access potable water. Purifying water is the first challenge, transporting it another and no less formidable second—and it is the latter problem targeted by the Q-Drum. Pieter Hendrikse comments: “In the early ’90s, my work took me through some of the rural areas and villages around Pietersburg and I noticed how the women and children struggled to get the water from the few taps to their homes, in some cases kilometers away. The few who had wheelbarrows used them, but mostly women carried the containers on their heads like in the rest of rural Africa, invariably causing many neck and spine injuries. I thought there had to be an easier way, and eventually me and my brother came up with the idea of the longitudinal shaft or doughnut hole through a cylindrical container. I made a prototype and the enthusiasm of the people in a nearby village was so great that we decided to go ahead with the project.”

❷ The form is that of a simple cylinder with a conspicuous hollow at the axis of rotation, referred to as the *longitudinal shaft*. While designed to be a purely functional form, a wheel reinvented, its elemental simplicity and roundness combine to create a friendly, cartoon-like affect. The body has but one opening, sealed with a threaded lid, which is positioned on a sidewall in contrasting color. It is the lid in conjunction with the longitudinal shaft that gives the drum the appearance of a *Q* in profile (assuming the drum is in the correct orientation). The drum is constructed of an inexpensive but durable linear, low-density polyethylene.

❸ The drum is transported by looping a rope through the longitudinal shaft and pulling. The longitudinal shaft is a robust, elegant solution—no handles or axles, no moving parts, and a rope can be easily repaired or replaced. Filled with 50 liters of water, the drum weighs approximately 55 kilograms, a rollable load manageable by even young children. The reduced effort required to move the Q-Drum enables children to be active helpers in an important domestic duty, potentially freeing women of the responsibility. The longitudinal shaft serves as a grippable recess for picking up and carrying as well for stacking, supporting filled stacked containers up to forty drums high. The shape affords itself nicely to makeshift application—for example, it is not difficult to conceive daisy-chained Q-Drums being pulled by a donkey or behind a cart, some filled with water, some with fuel, and some with foodstuffs.

❹ Wall thickness is 4 mm, sufficient to resist penetration at towing speeds on flat surfaces, but likely insufficient to resist penetration at runaway speed going down a hill. This is also the one real safety hazard of the product when used on steep inclines—you would not want to be downhill of a filled Q-Drum set free by a broken towline. A filled drum purportedly survives a three-meter drop-test and has been extensively used in rural areas of South Africa and Angola with good results, though for both practical use and safety reasons it is clearly best suited for relatively flat versus hilly regions, and even versus uneven ground.

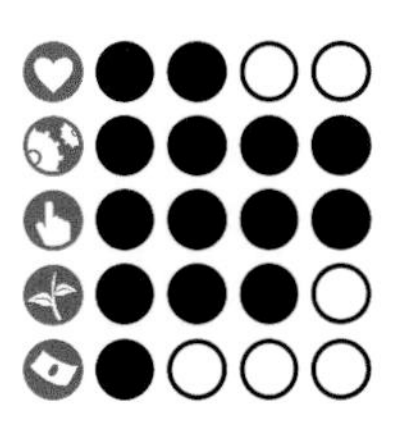

RAZR V3 Phone

Motorola, 2004

❶ A clear descendant of Motorola's StarTAC, the RAZR reintroduced the clamshell design to a market dominated by candy bar form factors. The phone soon became as much fashion statement as communication device, and its unique design reestablished Motorola as a leader in the industry. The question, however, is whether the RAZR was a fluke, the result of an internal "skunk works" team that cannot be reproduced, or a signal that Motorola was becoming a company, like Apple, equally adept at engineering and design. Jim Wicks, director of Motorola's Consumer Experience Design, comments: "A product by itself can't change a company. But, a product and a success can change a company. A product can start the change and can make people feel a part of that change. In the case of RAZR, I think the product alone has allowed us to gain much more confidence in a lot of things that we're thinking. However, you could also create a product that succeeds by accident and not realize it."

❷ The keypad is flat, cut from a single piece of nickle-plated copper alloy, enabling the wafer-thin form factor. The basic mechanism is reminiscent of the membrane keyboards of early computers such as the Sinclair and Atari. Click feedback is clear and definitive. Its symbols and grouping lines are achieved through cutouts as opposed to paint or decals, adding to the high-value sculptural quality of the phone. The grouping cuts, while functional, are overdone and clutter the display—some line subtraction is called for. The keypad is backlit, increasing legibility in dim light and reinforcing its Star Trek aesthetic.

❸ The software is slow and counterintuitive. Usability tests conducted by a variety of independent groups consistently reported 50 percent failure rates on task completion, and customer satisfaction surveys indicated that about 75 percent of RAZR users would not buy another Motorola handset, due to poor usability. These problems made the RAZR susceptible to competitors, like RIM and Apple. History tells us that the RAZR is not Motorola's first hit without an encore. Ron Garriques, head of Motorola's Mobile Devices, comments: "Nobody [at Motorola] took the StarTAC platform and turned it into a 'candy bar' or an email device. If we had, RIM wouldn't be alive today." The lesson here regards failing to understand deep user requirements—confusing form factor for product design, and then having the mistake unfortunately reinforced by a dramatic initial success.

❹ There can be little doubt that Star Trek has had more than its share of influence at Motorola. While the company's very Starfleet-like logo can be fairly said to predate the television series, having been created in 1955, their mobile phones going back to the MicroTAC appear to be increasingly refined approximations of vintage Star Trek communicators. The RAZR, however, does more than approximate—it is the Star Trek communicator, or at least what it would look like if created by a set designer today: flip style, ultra-thin, aluminum and magnesium case, concealed antenna, and a backlit, etched metallic keypad. At a time when mobile phones still looked very twentieth century, the RAZR let consumers warp ahead and buy a phone from the twenty-third century.

❺ The external display permits the device to report relevant information (e.g., who is calling), without opening the phone, essential for a clamshell configuration. The camera lens is appropriately centered directly above the display.

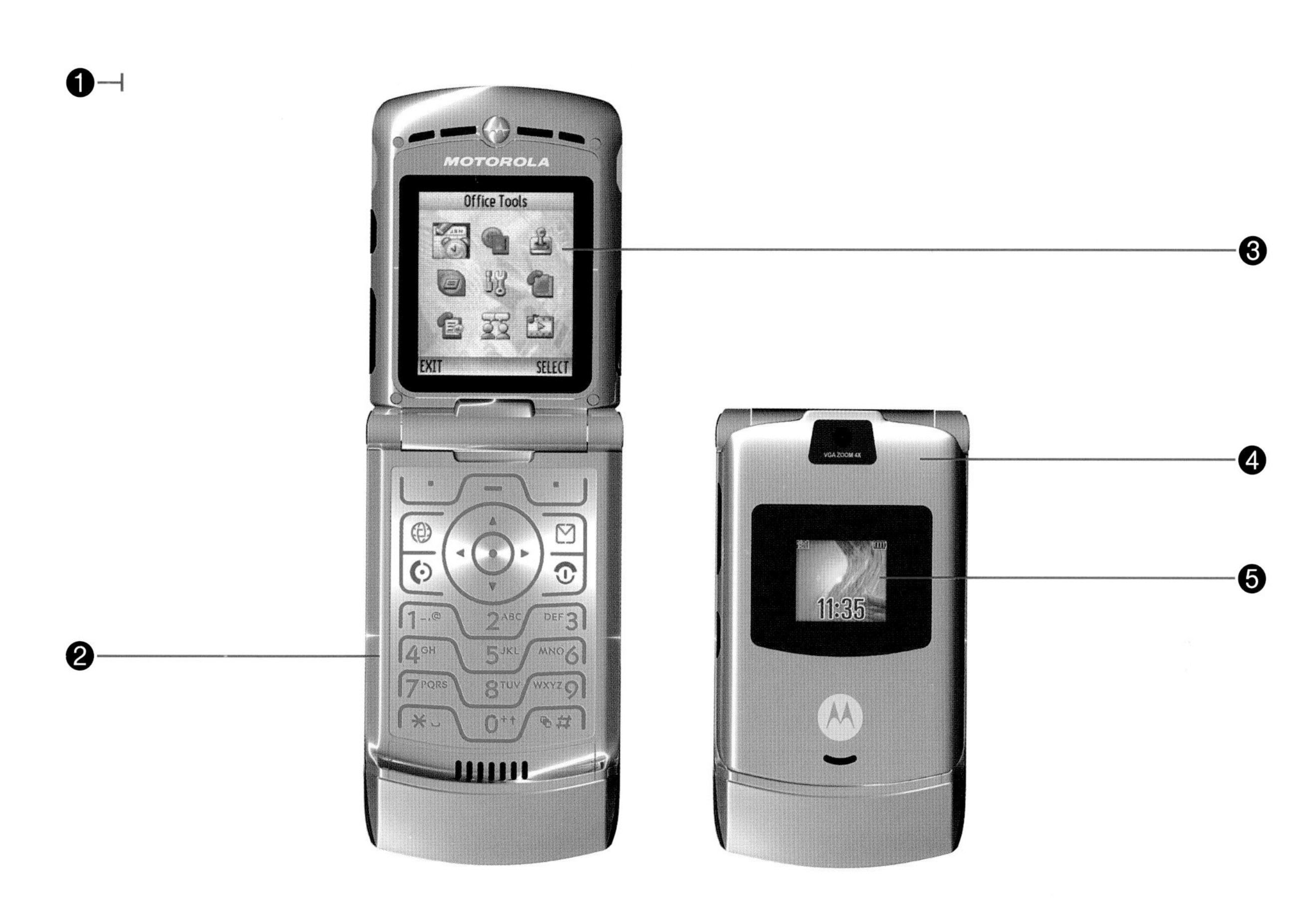

1
MOTOROLA
Office Tools
EXIT
SELECT
3
VGA ZOOM 4X
4
11:35
5
1
2ABC
DEF3
4GH
5JKL
MNO6
7PQRS
8TUV
WXYZ9
0
2

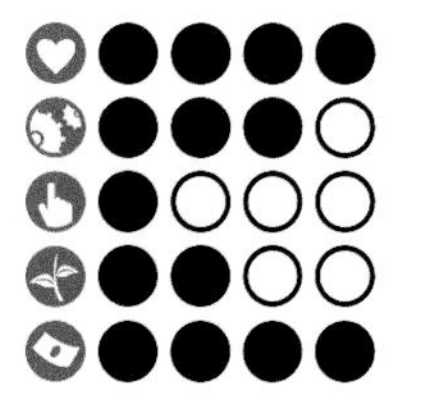

Remote Control — Series 1

TiVo, 1998

❶ With its distinctive organic shape, colorful keys, and intuitive on-television interface, the TiVo remote control has become the usability standard by which other remote controls are judged. While VCRs and DVRs around the world continue flashing *12:00* because users can't figure out how to set the time, TiVo users are happily browsing their online guides, recording programs, and speeding through commercials. Paul Newby, senior director of consumer engineering at TiVo, comments: "It became obvious early on that to achieve the seamless trick play and control we were after for our new PVR/DVR creation, the remote must be comfortable for long periods of in-hand use. This and an iconic look were motivations for the more distinctive, organic, peanut shape."

❷ Its sculptural shape, smooth curves, and beveled edges make one-handed use easy and comfortable. In terms of form, it is arguably the first remote control to properly rest in the human hand. The control resembles a symmetrical peanut, dog bone, or to some a phallus, giving it a unique aesthetic, but making it difficult to determine top-bottom orientation in a dark room. A similar problem relates to the dark gray color of the plastic body, which effectively camouflages the control in dim light and between dark sofa cushions. Flats have been added to the bottom to give it a stable resting position. The center of the body narrows and slopes inward toward the back, arching off of resting surfaces and making the control easy to pick up. The battery-bay cover is easily discovered, and slides to a close from the lower bottom of the control. The cover conforms to curve of the body and widens at the bottom, mirroring the shape of the infrared transmitter at the top to maintain symmetry.

❸ The Home button is properly placed at the top of the control, bearing the TiVo character logo and hosting an acrylic finish, the only button to do so. As the character is prominent in the TiVo branding, startup sequence, and menus, placing the character on the Home button effectively connects the user with the remote control, and both with the TiVo system.

❹ The "thumbs down" and "thumbs up" buttons are inconsistent with the rest of the button set — different icon style, palette, and label contrast — and garner a level of attention not commensurate with their import. They appear as artifacts of a design process that failed to edit them out.

❺ The buttons are colored by function and distinctively shaped, permitting what Newby has described as "Braille-ability." Tactile feedback is soft, grippy, and distinctive. Buttons are well grouped, based primarily on frequency of use, with high-use functions being located near the middle (the natural home position of the thumb), and low-use functions located at the top and bottom, based respectively on order of operation. Spacing between the upper and middle groups is poor, however, and the overall composition on the face appears jumbled.

❻ The Pause button is appropriately positioned, but as with the thumbs buttons, is overhighlighted. Highlighting and contrast are good, but cohesion with the overall design is also good, and the latter was sacrificed for canary yellow.

❼ The numeric keypad is the bottom-most grouping and arranged in squared rows and columns, instead of a distribution that follows the lines of the body and fills the widening space, a mistake corrected on Series 3 controls.

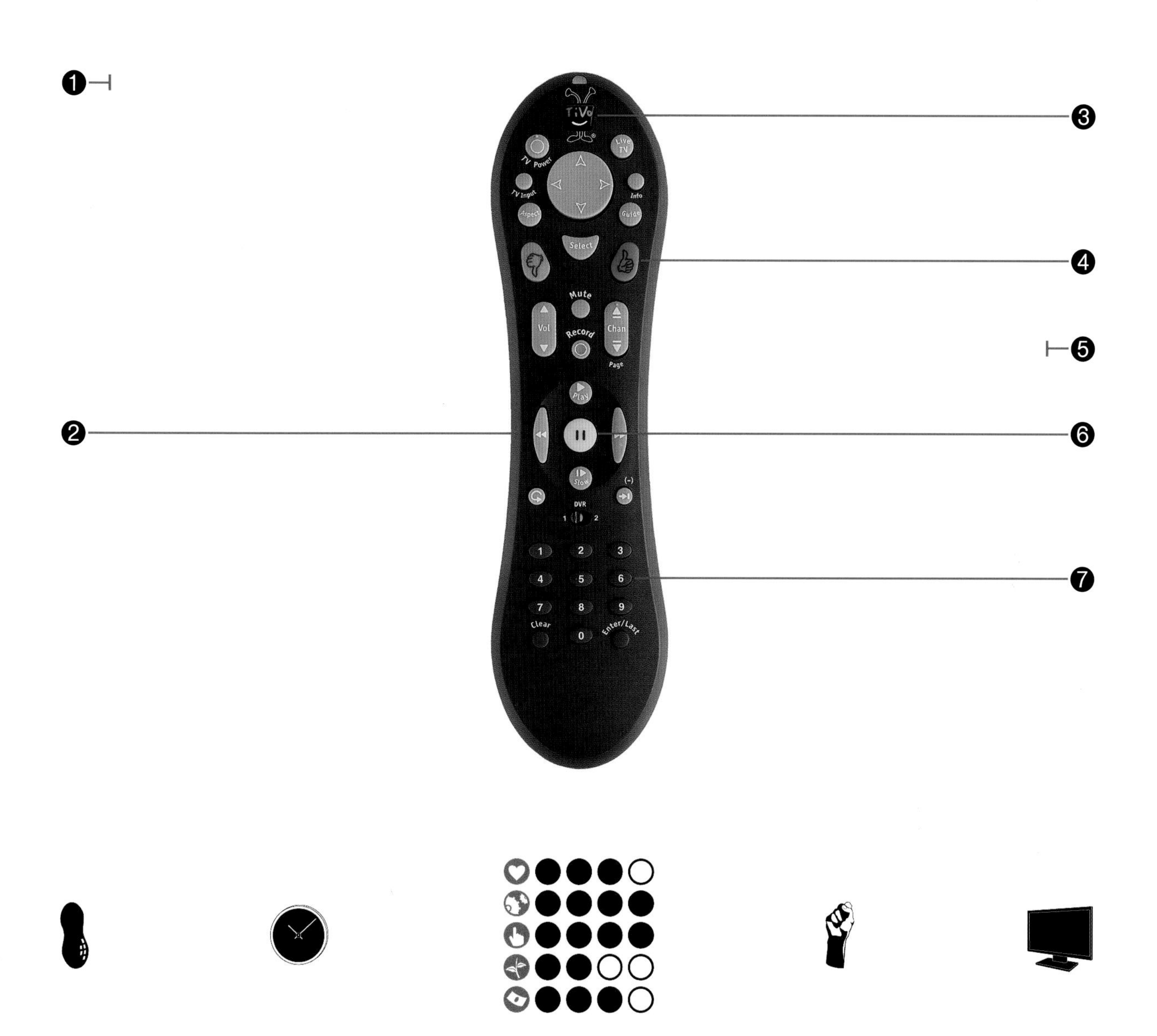
1
2
3
4
5
6
7
TiVo
Live TV
TV Power
TV Input
Info
Aspect
Guide
Select
Mute
Vol
Chan
Record
Page
Play
Slow
(-)
DVR
1 2
1 2 3
4 5 6
7 8 9
Clear
0
Enter/Last

Roomba Robotic Vacuum Cleaner

Colin Angle, Rodney Brooks, and Helen Greiner, 2002

❶ That robots should ultimately appear anthropomorphic or zoomorphic has always been a largely unquestioned assumption among robot enthusiasts—consider early examples such as the musicians of Al-Jazari, the roboknight of Leonardo, or more modern examples such as AIBO, Pleo, and Kismet. Then, with minimal fanfare, along comes the Roomba, an unassuming disk-shaped robot bearing no physical resemblance to anything biological. Roomba not only becomes the first commercially successful consumer robot, it unexpectedly becomes a robot with which people develop strong emotional connections. Helen Greiner comments: "Our customer base—homemakers, people who just want to get the vacuuming job done, people who are used to doing it themselves—buy the Roomba as an appliance, but once they get it home, it's going around, it's following the wall, it's getting around the furniture, it doesn't fall down the stairs, it has bleeps and bloops to communicate. The only thing in their experience that has acted that way has been a pet. So people actually start to name it. You don't see anyone name their toasters, but a lot of people tell me they have named their Roomba."

❷ When stationary, Roomba's affect is flat. However, when the robot is activated and begins zipping across the floor sucking up dirt and dust, this changes—it becomes industrious, determined. Roomba illustrates two important lessons for designers: 1. Ambiguous forms, free of biases related to superficial resemblances, establish emotional connections based primarily on function and utility; and 2. Familiar forms often set high expectations by analogy that can be difficult to match (e.g., AIBO versus a real dog), whereas ambiguous forms let expectations develop empirically through experimentation and use.

❸ Rather than complex sensing and wayfinding intelligence, Roomba uses a series of basic algorithms to clean—wall-following, spiral cleaning, angle changing after contact with a barrier—a signature of Brooks' philosophy that robots should be like insects, guided by simple control mechanisms tuned to their environment. A removable dustbin yields evidence that the robot, which at times seems to swirl aimlessly and needlessly revisit well-vacuumed regions, is actually on task removing dirt from carpet. While this random-walk approach scores high in terms of coverage and cleaning, it can often lead to the Roomba getting hung up on furniture or scattered objects. Accordingly, rooms must be Roomba-proofed before they can be cleaned. In any other context this would be considered a severe inconvenience, but research by Ja-Young Sung et al. actually found that users often perceived this pre-vacuuming ritual in a positive context, as doing their part since the robot was doing the really hard work. Like Jibbitz for Crocs footwear, third-party accessories for the Roomba (e.g., furry pullover covers) have become available to let users personalize their robot.

❹ Key to Roomba's appeal seems to be its relative autonomy. Unlike social robots (e.g., Pleo and AIBO) that require close and persistent attention, the Roomba can essentially be left alone once a room has been prepared. This autonomy makes the Roomba more like an independent contributor to the household, and less like a dependent child or pet in need of constant attention and oversight. There is also an air of honesty with the Roomba that is missing with social robots—no feigned expressions or simulated emotions, just a humble bot focused on keeping the floor clean.

1
2
3
4
iRobot
Roomba
Intelligent FloorVac
S
M
L
Select Room Size to Start / Stop

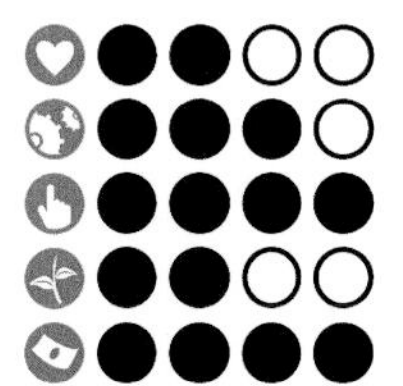

Round Thermostat

Henry Dreyfuss Associates and Carl Kronmiller for Honeywell, 1953

❶ A critical but rarely discussed aspect of real-world product design is persuasion. Great product designers have historically been great pitchmen—from Raymond Lowey to Philippe Starck—able to excite and educate clients about design, which then empower the client leadership to bring their revolutionary products to market. Henry Dreyfuss was no exception to this rule, and it was likely his influence with Harold W. Sweatt, president of Honeywell, that actualized a project Dreyfuss had been thinking about since the introduction of his round Big Ben clocks in 1939: a round thermostat. He observed that rectangular thermostats, which were the only form factor available at the time, always appeared crooked when mounted on walls. A round thermostat would solve this problem. Dreyfuss, known for his napkin-based doodling during conversations—in particular, he enjoyed practicing drawing perfect circles—began discussing the possibility of a round thermostat with Sweatt in 1940. Later that same year, in a discussion between Sweatt and Honeywell engineer Carl Kronmiller, the seed for developing a round thermostat would be sown. Carl Hoyt, Honeywell employee, comments: "H.W. [Harold W. Sweatt] came to [Carl Kronmiller's] area one day in 1940, sat down at his desk, and they talked about products. Carl told H.W., 'We need a radically new thermostat. Why can't we make something a lot different from our competitors' models and different from our own thermostats?' Carl told me as they talked, H.W. picked up a piece of paper and started drawing circles on it. Then he handed the paper to Carl and said, 'Here. Go ahead and make something of it.'" The seed would take over a decade to bear fruit, but the round thermostat would come to define Honeywell product design for the next fifty years.

❷ The product, marketed as the "Honeywell Round," was an aesthetic leap over the staid rectilinear thermostats of the time. Its small round form appealed to women, and not only did it satisfy Dreyfuss' issue of never appearing incorrectly oriented, the cover of the thermostat could be easily removed and painted to blend with the wall. It was the first thermostat to consider the interior design of the home.

❸ The red arrow indicates the current temperature. The black arrow, which is fixed to the transparent plastic dial, is rotated to the desired temperature. The dial is the only movable element, and its serrated perimeter clearly affords turning. The display is less refined, as it is not immediately evident what the red and black arrows mean—which is the setting and which is the temperature? It is interesting to note that Dreyfuss' patent sketches represent temperature using a somewhat familiar crescent-shaped mercury thermometer, eliminating the confusion by representing room temperature and the temperature setting in visually distinct ways.

❹ Temperature is roughly marked in two-degree increments because the accuracy of the thermostat's spiral wound thermometer was poor. This low fidelity created problems when the black arrow was precisely aligned with the red arrow, because the actual temperature could be plus or minus several degrees—the represented accuracy of the thermostat did not match the functional accuracy of the thermostat, and the product was often perceived to be defective as a result. Later models moved the thermometer to the bottom of the dial, such that the temperature-setting scale was on top and the thermometer scale was on the bottom. This change eliminated the possibility of ever seeing the arrows in precise alignment, and correspondingly eliminated the reports of the thermostat's being defective.

❶

❷

❸

❹

70

60 80

50 90

HONEYWELL

Honeywell

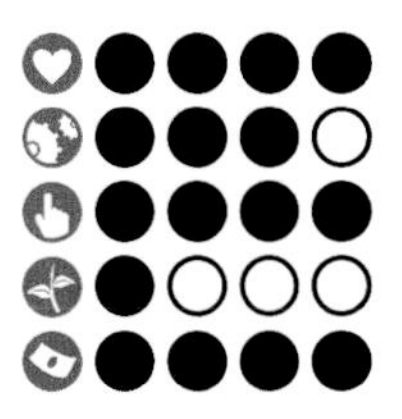

Rubik's Cube

Erno Rubik, 1974

❶ Few puzzles have managed to combine the formidable difficulty (just over forty-three quintillion possible combinations) with the *prima facie* simplicity (one solution no more than twenty-five rotations away from any given configuration) of Rubik's Cube, which may explain why it has become one of the best-selling toys of all time. More than twehty-five years after its introduction to the West, the original Cube and numerous variations continue to sell well, and Rubik's Cube solving competitions, called *speedcubing competitions*, are increasing in popularity. Of this writing, the world record to solve the Cube is 7.08 seconds, set by Erik Akkersdijk at the Czech Open 2008. How does one go about creating such a successful product? Erno Rubik comments: "I did not plan to make the Cube. I did not plan the success. I wanted nothing else than to make the object as perfect as possible. Now, after the Cube, I still don't have any plans to make anything like it. I'm still the same person, thinking the same way, so it's possible I will invent something. But to want to repeat the Cube—that is not the way to live."

❷ The form is a simple cube comprised of twenty-six smaller cubes or cubelets, configured as a 3x3 grid on each of the six faces. Cubelets are affixed to a central mechanism, an engineering puzzle in its own right, which allows each grid row and column to be rotated independently. The object is solid, easily held, and its manipulation easily discovered—for example, there are documented cases of very young children, as young as three, easily using and solving the puzzle.

❸ Unlike many puzzles, Rubik's Cube arrives in the solved configuration. The arranged state promotes a calm and peaceful affect, appealing to innate desires for order and security. With just a few twists and turns, however, order is long lost—like entering the Labyrinth of Crete without the ball of yarn to trace your way out. Backtracking is soon deemed futile, and the focus turns to just getting a single side in the correct configuration. But each twist that attains order on one face compromises it on another. Play turns serious, frustrating, and often addictive. Rubik has said, "We turn the Cube, and it twists us." Why is the need to restore order in the Cube so compelling? Several psychological principles are identifiable: people naturally prefer symmetry over asymmetry, order over disorder, and complete over incomplete cognitive tasks. When closure is not attained, the brain continues to work on the problem, a phenomenon known as the Zeigarnik effect. History is also instructive. Rubik's Cube is reminiscent of the classic 15 Puzzle devised in the late 1870s, itself the cause of a puzzle mania second only to Rubik's Cube. A 15 Puzzle consists of fifteen consecutively numbered tiles that can be pushed around inside a square frame. It shares many of the qualities of the Cube: it arrives in an arranged state, it's easy to hold and manipulate, order is easily lost and difficult to restore, and it presents a façade of simplicity, concealing the true complexity of the task and leading users to think the puzzle is something that they can solve. The parallels are difficult to ignore, and no doubt suggest some of the keys to their respective successes.

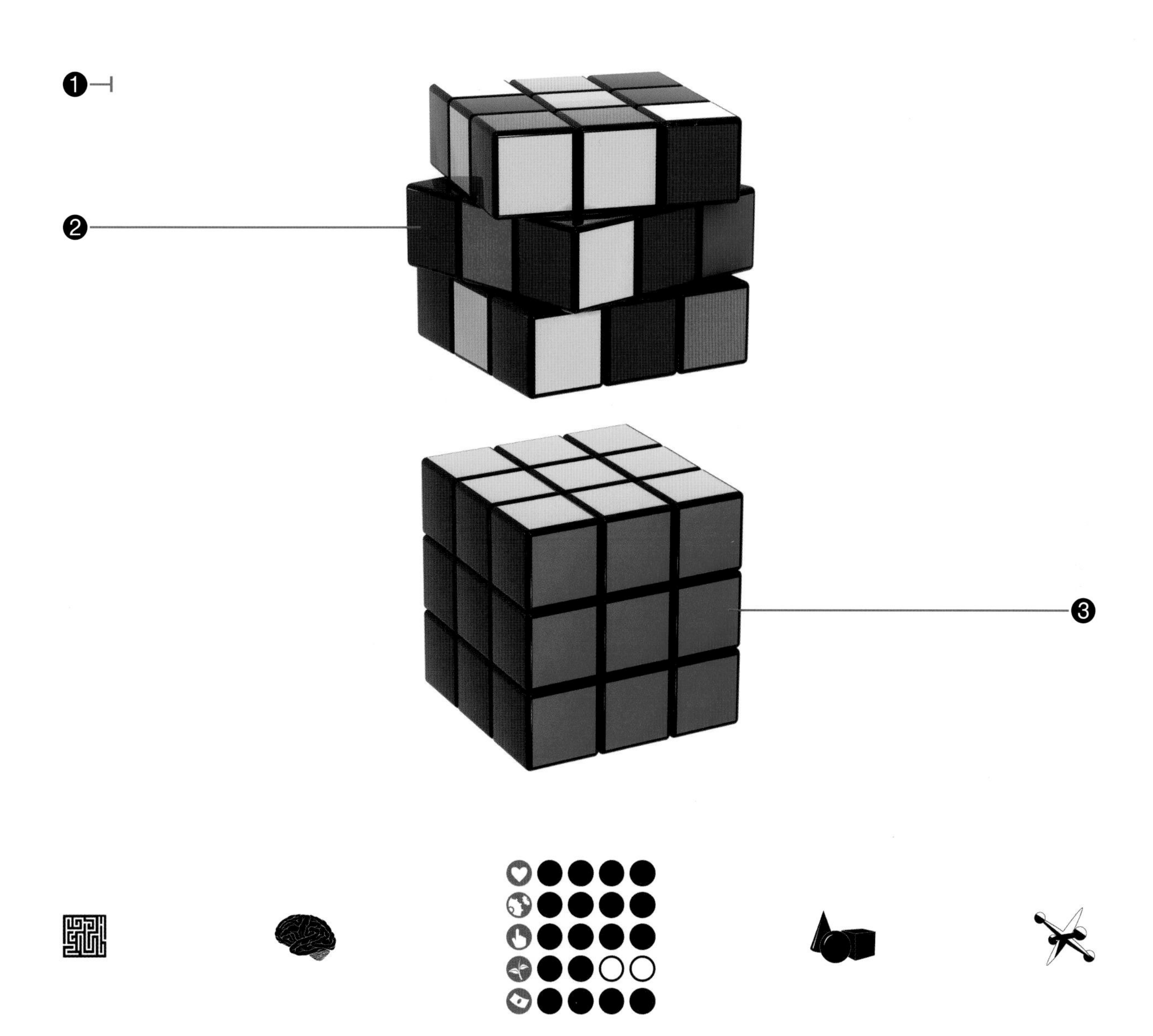

Safety Razor

King Camp Gillette and William Nickerson, 1992

❶ In 1895, King Camp Gillette worked as a salesman for William Painter, the man who invented cork-lined bottle caps. Painter advised Gillette, "You are always thinking and inventing something. Why don't you try to think of something like the Crown Cork, which, when once used, is thrown away, and the customer keeps coming back for more—and with every additional customer you get, you are building a foundation of profit." So began Gillette's hard-target search for needs that could be satisfied using Painter's single-use strategy. Then, one morning, while shaving with a dull razor, Gillette had an epiphany that would ultimately lead to the disposable razor—and, more significantly, to a "disposable society." King Gillette comments: "As I stood there with the razor in my hand, my eyes resting on it as lightly as a bird settling down on its nest—the Gillette razor was born. I saw it all in a moment, and in that same moment many unvoiced questions were asked and answered more with the rapidity of a dream than by the slow process of reasoning... I stood there before that mirror in a trance of joy at what I saw."

❷ Although the safety razor had been invented in the 1870s, they were still quite rare and expensive. Straight razors, also known as "cut-throat" razors, and barbershop shaves were still the norm. Gillette's innovation was to make money selling inexpensive razor blades with short life cycles (approximately ten cuts, though twenty to forty were advertised), enabling the razor itself to be given away or sold at a loss—a business strategy well known today as the *loss-leader* model. Year one sales: 51 razors and 168 blades. Year two sales: 91,000 razors and over 2 million blades.

❸ The aesthetic is basic and simple—a reminder that product evolution, like biological evolution, often does not confer progress or improvement with time. The quality of the shave is quite good. So good, in fact, that many who take their shaving seriously often seek out early twentieth-century razors in lieu of their modern blinged-out, multibladed counterparts. The razor is comprised of three parts: handle, curved lower base plate with safety guard, and a curved upper compression plate. The razor blades are inserted between the two curved head plates, and then the entire head assembly screws into the knurled handle. As the head assembly is torqued down, the cutting edges of the razor blade angle downward to match the curvature of the plates—this in combination with the toothed safety guard is what makes the razor "safe." Replacing razor blades, however, is less safe. Although disassembly is simple, the thinness and sharpness of a double-edged razor blade makes handling a perilous activity.

❹ The actual make-it-work design and engineering of both the razor and the blades came from William Nickerson, a brilliant inventor whom Gillette commissioned to solve the key problem of his razor: manufacturing thin, sharp, and cheap blades that would obsolesce at a predictable rate. "The problem," Nickerson said, "is entirely different from that involved in the tempering and grinding of ordinary razors and other keen tools, not only on account of the thinness of the blades, but also on account of the cheapness with which it must be done." Nickerson would ultimately solve these problems and others, and though he desired equal public recognition for his work, his unfortunate last name did not lend itself to selling razors. The company name would belong to King Gillette alone.

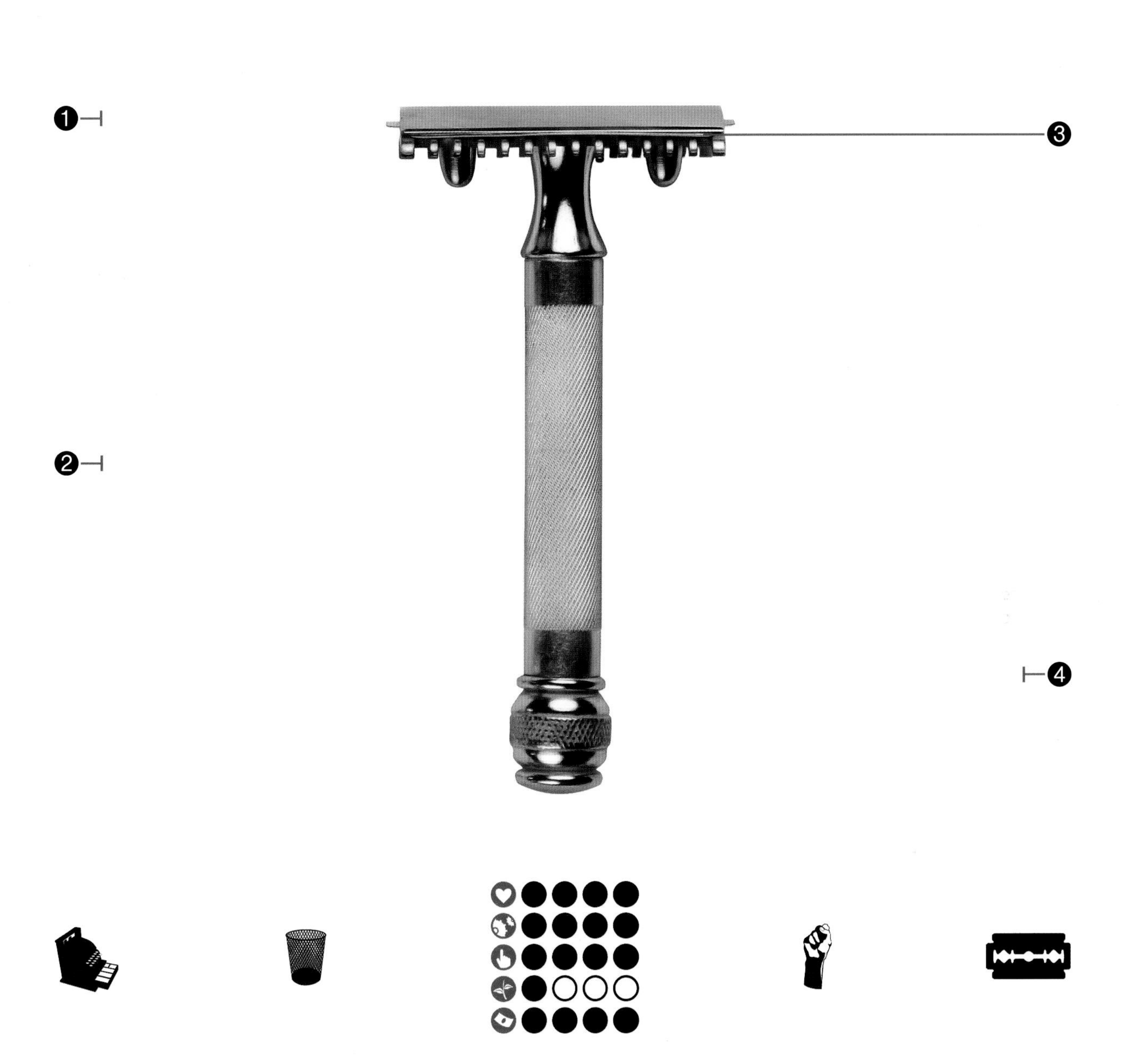

Segway i2

DEKA Research and Development and Segway, 2006

❶ Since its much-hyped introduction in 2001, the Segway has become an object of desire to the geek chic and an object of ridicule to just about everyone else. Critics assert that the product is more an homage to engineering than a practical solution to a real problem—a modern-day South Pointing Chariot. Proponents argue that change takes time, and that the economic and environmental forces at play in urban regions make a solution like the Segway inevitable. Dean Kamen comments: "Over 50 percent of the car trips in the United States are less than three miles. The average speed within city limits between any two points is less than 9 mph. Then why does everybody use his or her car to get around? Because walking is less than 2 mph! What if they could get on a Segway and cruise from start to finish at 8 mph? That's the same speed at which a taxi travels. And, it's cost effective, energy efficient, environmentally friendly, and fun. In highly dense urbanized areas where buses and cars do not work well—they have only been used to fill the gap for lack of a good alternative—the Segway shines."

❷ The Segway has always suffered from a kind of toy scooter-motorcycle identity crisis, and this model, though improved, is no exception. The form itself is intuitive, leaving no doubt where prospective riders are to put their feet and hands, but the affect is utilitarian. If only it had the personality of its logo. The i2 is available in three anodized finishes—black, white, and gold—a retreat from the more colorful palette of second generation models, and appropriate to support its positioning as a high-value product.

❸ The handlebars on the i2 are much refined over the pullback bars of the original HT, cropped and straightened to look less touring-bicycle and more motorcycle-dragster. The bars sit atop the new LeanSteer frame, the most significant design advance in this model. To turn left or right, the rider now simply moves the LeanSteer monobar frame left or right. Not only does this make turning more intuitive than its predecessors, it requires riders to move their arms and bend their knees to lean into turns—riders of this model look more like skiers swaying and flexing into turns than the hapless free-falling passengers of previous models. It is a solid step toward redefining the Segway as a vehicle for recreation versus functional transport, a vehicle for the active and healthy versus the passive and infirm.

❹ The greatest challenge for the Segway—aside from its high price point—is the exposed sedentary and awkwardly upright posture of the rider, which feeds a negative perception of the product: bicycles are associated with fitness, rollerblades with recreation, Segways with sloth. Harley Davidson, too, has exposed and sedentary riders, but any perception of their riders as "slothful" is overwhelmed by their design emphasis on power and dominance—bigger, faster, louder—an image reinforced by marketing their riders as archetypal outlaws. These strategies are effective—consider the perception of scooter riders by comparison. It should be noted that with the introduction of LeanSteer, the i2 requires significantly more visible rider activity than do earlier models. Progress, but still needs more Harley.

❺ Wheel design is banal, and given their size and weight in the design, represents a missed opportunity to communicate a compelling product story.

1
2
3
4
5

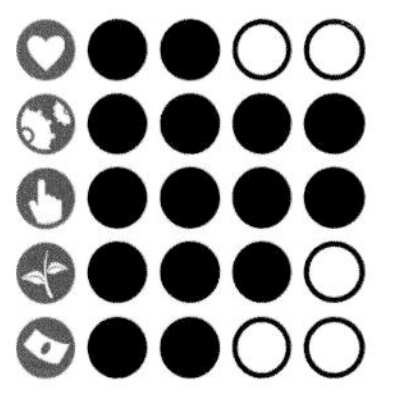

Selectric Typewriter

Eliot Noyes, Bud Beattie, and Allan McCroskery, 1961

❶ Soon after Eliot Noyes began working with IBM on a general design program, his firm was given the opportunity to evaluate the typewriter line of products. Noyes heard about an internal project inspired by a toy typewriter, which led engineers to revisit the classic Blickensderfer, an innovative typewriter invented in 1893 with a removable typewheel. Unfortunately, the project had stalled because the cylindrical shape of the typewheel had problems delivering good-quality type impressions. After much experimentation and research, a solution to the problem was discovered—a solution appropriately inspired by a lightbulb. Allan McCroskery, designer at Eliot Noyes & Associates, comments: "They were going nuts how to work this thing out. There was ... a development engineer, Bud Beattie, and it was his job to get this thing to work. He went home and his wife was complaining about a light bulb that had blown. He screwed the light bulb and, like in the cartoons, the light went on. He quickly got a grease pencil, drew circles [on the bulb], and divided it up. This is it!"

❷ The Selectric typewriter was the first office product to advance the notion that office equipment should be attractive. Believing that a sculptural form would give the typewriter a timeless quality, Noyes worked with McCroskery to shape the typewriter body in the spirit of Isamu Noguchi, a sculpture by the sculptor who had provided art for IBM's headquarters. The Selectric was offered in a variety of bright colors, as well as custom colors—for example, Selectrics were sold to the University of Kentucky in Wildcat Blue. Combine curved lines, roundy edges, and a colorful palette, and you get a form factor tuned for women—who were, of course, the principal users of the product.

❸ Unlike a traditional typewriter, the carriage of the Selectric does not move when typing—minimizing its required desktop footprint—nor are there type bars to strike the paper. The Selectric uses a golf ball–size typing element called a "typeball" that moves along the fixed platen with each key press, tilting and rotating to deliver the typed character, number, or symbol through an inked ribbon. The typeball can be easily replaced with others bearing different typefaces, a significant step forward in word processing. Key response is much faster than type bars. In the event that the user can type faster than the electronics can handle, the Selectric is equipped with a storage system that holds the second letter in common combinations until it can be printed. The inked ribbons are enclosed in plastic cartridges that snap into place, requiring no threading and creating no mess. Use, like the aesthetic, is clean and simple. The Selectric line proved Noyes' contention that "good design is good business," capturing 75 percent of the U.S. typewriter market in just fifteen years.

❹ The Selectric keyboard is cut into a recessed region that looks scooped out, framing the keyboard and containing the hands without breaking the continuity of the lines. The curve extends downward from the back, giving the keyboard a slightly dished profile—an ergonomically superior profile to either a flat or stepped configuration in terms of typing and error rates. The keyboard retained a QWERTY layout, which was invented in 1878 to prevent the clashing and binding of type bars that would occur with rapid typing. Although the speed and mechanical refinements of the Selectric (and earlier typewriters) make QWERTY unnecessary, the standard persists despite higher-performing layouts. Lesson: convention often trumps performance.

1
2
3
4
IBM

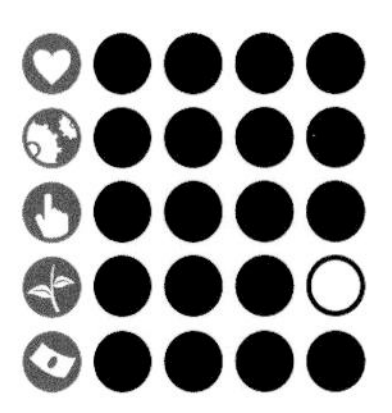

Silent Cello SVC-200

Yamaha, 2002

❶ Replicating the performance of a long-standing product through a different means Is a design strategy that can dramatically expand a performance envelope and address otherwise intractable problems intrinsic to the traditional product design. Such was the case with molded plywood chairs, microwave ovens, and digital cameras, and such is the case with the Silent Cello. Shigeto Nomura, general manager of strings, guitar, and percussion exports for Yamaha Corporation of Japan, comments: "[Professional cellists] needed an instrument to practice at a hotel or at midnight, and they also saw great potential for big stage use with other electric instruments ... Acoustically, we used natural woods, spruce—the same as an acoustic cello—as the centerpiece. The instrument's slim body made the natural sound harsher, but we successfully adjusted the warmer sound by modifying an electrical circuit ... Our wide experience with designing and producing many acoustic and electric instruments helps make such hybrid products. Another factor is our relationship with musicians. Many of our employees are musicians, and sharing ideas and passions with world-class artists is the major driving power in developing these instruments."

❷ The form is sparse, but recognizable as a cello—at least to cellists. The instrument retains the distinctive top scroll, four tuning pegs, and general centerline equivalence down to the spike at the bottom, but the rest of the body is modern sculpture. The shoulder rest, hand rest, and knee supports require some imagination to fill the form, but Gestalting the lines is a satisfying experience and gives the cello an intriguing aesthetic. Electronics are concealed in the back of the body, the most important feature of which is a headphone jack that allows a rich auditory experience for the user and significantly muted play to the outside world.

❸ As a practice instrument, the silent cello is a tradeoff. Although it is undeniably more portable and usable in more contexts because of its silent-play mode, there are genuine differences between it and a traditional cello that could compromise technique. For example, the Silent Cello requires a much lighter touch and is less forgiving of string errors. Clip the wrong string with the bow or prepare strings poorly on the fingerboard, and the shrieks and squeaks are immediate, unlike a traditional cello in which the play energy has to achieve a minimum threshold to drive through the big body of the instrument. Additionally, SVC-200 does not produce resonances or "wolf tones"—that is, when a played note matches the natural resonating frequency of the body of a traditional cello, it produces a sound akin to the howling of a wolf. Response is also artificially even and sustains last much longer than with a traditional cello. As with an electric guitar, the silent cello is more interesting and effective as a stand-alone instrument than as a practice substitute for its acoustic counterpart.

❹ The corniced knee supports are adjustable for comfort and collapsible for transport, and provide the essential lower contact points required by the player. In a traditional wooden cello, cornices add necessary structure to the body. Here, however, they are ornamental. Perhaps the sense among the designers was that so much had been cut away that some semblance of traditional form was necessary to give the instrument a recognizable identity. Once you pass a Loewy "commando styling" threshold, however, the proverbial mitts need to come off. In a radical redesign, better to let the product boldly redefine itself, than to apologetically retain vestigial elements of classical forms.

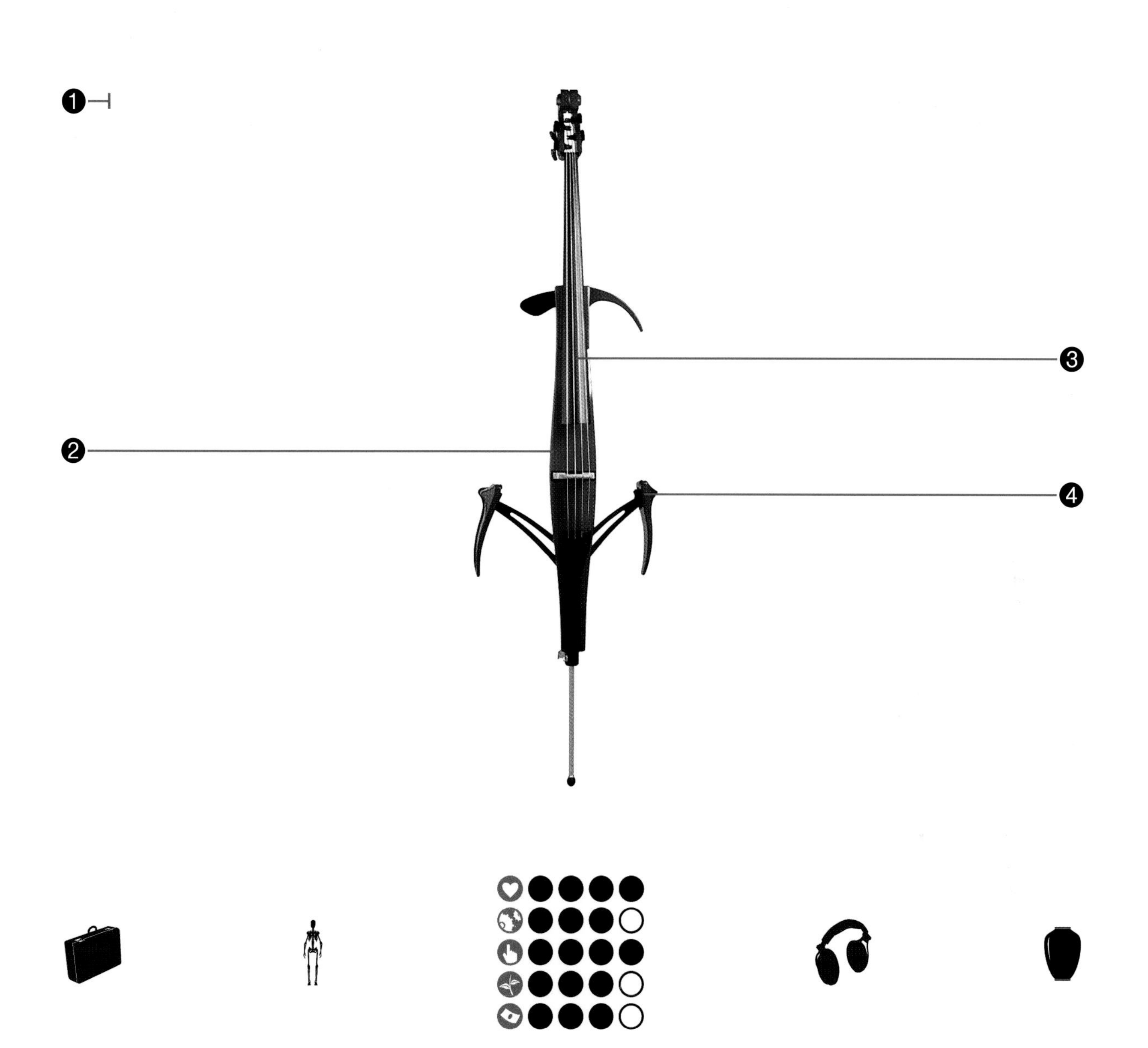
1
2
3
4

Sixtant Electric Shaver

Hans Gugelot and Gerd Alfred Müller for Braun, 1962

❶ Successor to the S-50 shaver introduced in the 1950s, the Sixtant by Braun—pronounced "brown," not "brawn," though the latter mispronunciation is common and serves the brand well with regards to male-oriented products—would become one icon of many born of a revolution in industrial design known as the Ulm movement, after the Ulm School of Design (HfG) in Ulm, Germany. The school extended Bauhaus principles to product design, resulting in a near pathological emphasis on aesthetic minimalism and the "form follows function" dictum. Dieter Rams, former chief of design at Braun, summarizes the philosophy as practiced at Braun: "By concentrating on the 'bare necessities,' by order and harmony, by renouncing all that is unimportant and superfluous we meet with products of high complexity. They are beyond all modern trends. They express the essence of design. Countless Braun models were produced—unaltered for years and years."

❷ The form is basic, masculine, sized and shaped to be held in one hand. The color choice was unique for hygienic products of that time, which were typically colored white to connote cleanliness. The notion that conservative color palettes increase the perception of value of products, especially those that employ neutrals and natural material finishes, was likely first formalized at HfG, and is certainly manifest in the Sixtant, which was priced as a premium shaver. The substantial black case affords gripping and effectively conceals dirt and wear. The contrast between the black body and polished metal head is striking, sophisticated, akin to an expensive automobile with a refined chrome grille. One can dislike the Ulm styling, but the timelessness of the design speaks to postmodernists—trying to teach them something.

❸ A key innovation of the shaver is its foil cutter, which made the notion of a "fast and dry" shave compatible with a "safe" shave. The mechanism is simple and essentially incapable of causing injury. The steel foil head enshrouds an oscillating cutter block. The foil flexes, allowing the razor to be pushed against the face, which isolates and tugs whiskers outstretched so that the vibrating cutter block can shear them off. The resulting shave is not as close as a razor, but still quite close and without the worries of nicks and cuts. The cut whiskers are mostly captured in the head, which is removable and can be dusted clean with an included brush. To one side of the foil is a sideburn and moustache trimmer, a useful and virtually invisible addition in the design. As an electric shaver, the Sixtant is more portable and convenient than a wet-shave razor. And, in contrast to a razor shave, where missed spots stand out quite noticeably against shaved areas, an electric shave is uniform and can be easily touched up if needed. Drawbacks include whisker dust that is not captured by the foil, which is not insubstantial, and a comparatively loud shaving experience—no shaving quietly in the early morning with a companion sleeping nearby. Additionally, the tugging motion of the foil often irritates those with sensitive skin, resulting in rashes and ingrown hairs. The basic mechanism of Braun electric shavers remains unchanged to this day, though the pure Ulm genes that once defined the Braun phenotype have long been suppressed, overwhelmed by marketeers and pop influences.

1
BRAUN
2
3

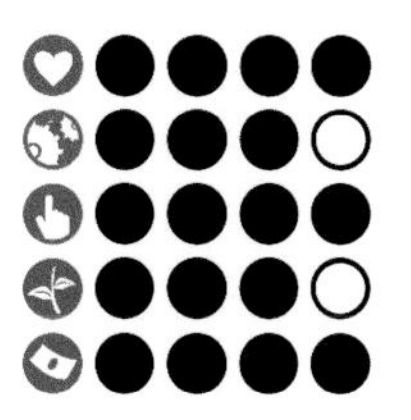

Soda King Siphon

Norman Bel Geddes, 1935

❶ Although he tends not to receive the contemporary recognition of peers such as Loewy, Teague, and Dreyfuss, one cannot deny that Norman Bel Geddes was a visionary designer. Admirers compare his mostly unrealized conceptual work to no less than that of Leonardo da Vinci, while detractors prefer a P.T. Barnum comparison, the latter due to his flamboyant manner and penchant for self-promotion. Either way, it is clear that Geddes was able to look beyond many of the accepted design and engineering constraints of the day — perhaps due to his background in art and stagecraft — and explore visionary solutions to difficult problems. Like da Vinci, his conceptual work was ambitious, exploring designs for utopian cities, vast interstate highways, and amphibian airliners. Given the large shadow cast by these visionary explorations, it is not surprising that much of Geddes's practical work is little known. This is a shame, because there is much to be learned from this work, like the guiding principles he used to design them. Norman Bel Geddes comments: "There are two cardinal principles by which the designer should always be guided: simplicity and the use of interesting materials. Simplicity means, of course, the avoidance of excess decoration and the elimination of every unnecessary detail. The designer who works in industrial fields is using, instead of pigment on canvas, or marble, a combination of materials that are the development of this age. They are as different from the materials of the painter as the materials of the painter are from the materials of the sculptor. In addition, of course, he has aesthetic problems to solve — proportion, for instance. There is a great difference between merely satisfying proportion and the proportioning of a form that gives it added interest, vitality, conviction and distinction."

❷ Like so much of Geddes' work, the product design is as much an exploration of the future as it is a soda dispenser. If a Soda King were transported back in time 5 million years, it is not hard to imagine that it might well lead to a tribe of hominids throwing sticks in the air. Considered Art Deco due to its chrome plating and unnecessarily low drag coefficient, the siphon is an early example of streamlining applied to everyday objects. The eccentric feature of the form is its triangular beak, which houses the pouring spout and insertion point for carbon dioxide cartridges. Given this context, it seems plausible that Geddes was modeling the spout after an engine exhaust or thruster port. Usability is excellent with one visible button on top of the body.

❸ The Soda King is propositionally rich, which means the number of propositions or messages expressed by its form factor is high. Propositional density does not necessarily make a form pleasing to the eyes, but it is what makes a form interesting to the brain. In the case of the Soda King, for example, its form could be a rocket, bullet, bird, alien, fire extinguisher, sculpture, or face, akin to an Easter Island totem. These varied impressions activate different memory centers in the brain, catalyzing comparisons and analogs, and evoking different and often subconscious emotional responses. The trick, of course, is to achieve high propositional density without high complexity. It is no great feat to communicate many messages with many elements, but it is an intriguing psychological play to communicate many messages with few elements — it is this phenomenon that makes puns funny and double entendres clever, and the design of Geddes Soda King siphon classic.

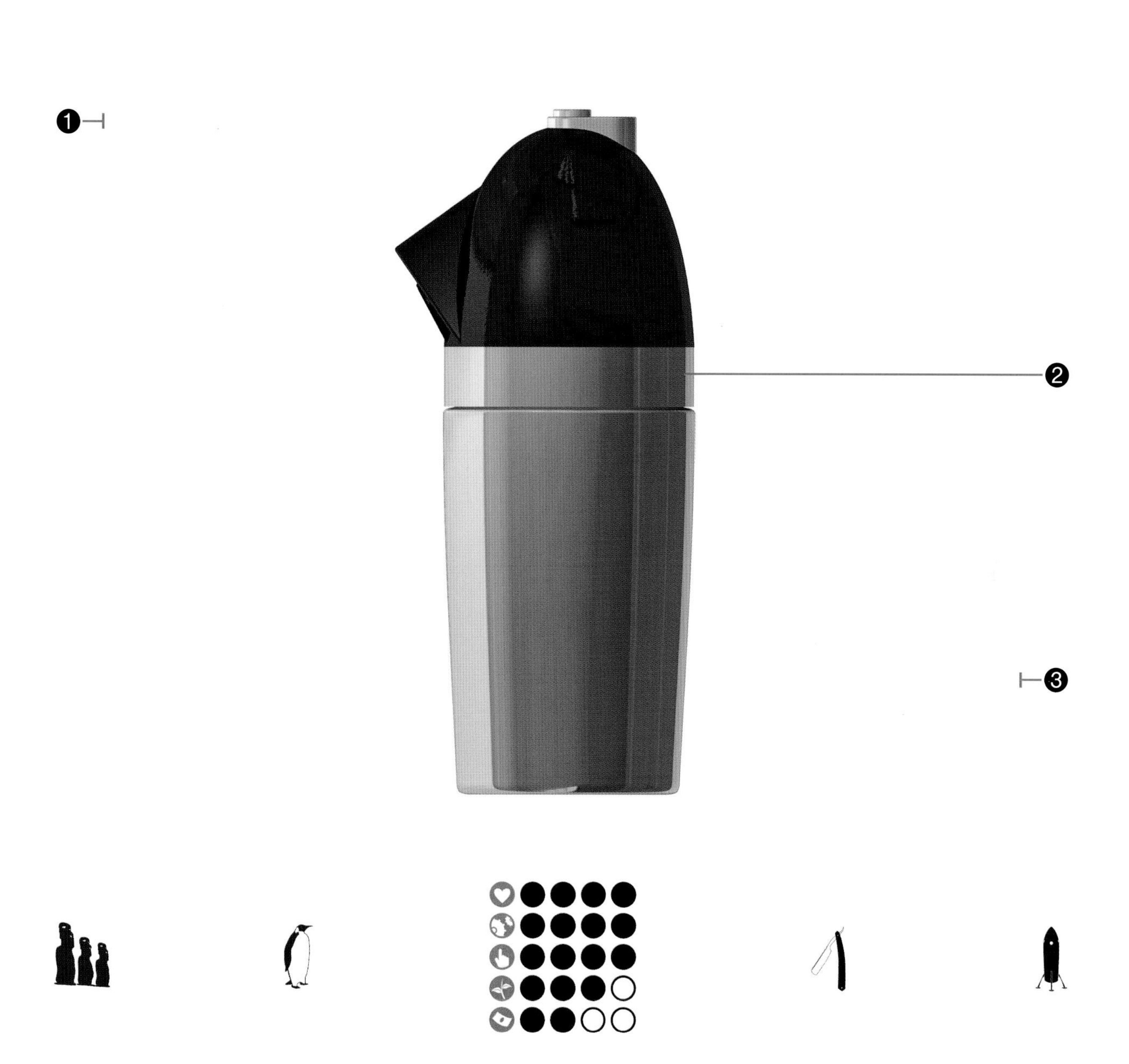

1
2
3

Sonicare Toothbrush

David Giuliani, David Engel, Roy Martin, and Steve Meginniss, 1992

❶ Having had a very personal and painful experience with periodontal disease a few years earlier, David Giuliani was excited to learn about research going on at the University of Washington that studied an improved method of cleaning teeth using sonic technology—a method that not only cleaned teeth better, but that dislodged dental plaque beneath the gums. Giuliani licensed the technology and worked with a small team of designers and engineers for over five years to transform the research technology into a commercially viable product. David Giuliani comments: "We designed the toothbrush to be simple to use. We wanted it to be so natural and so intuitive that there would be no need for an operator manual—no need to translate anything into twelve different languages. So easy that kids could just pick it up and use it. So solid and robust that you could drop it on the floor like a brick and it would continue to function. This is why we used a sealed-case design, with no shafts or exposed mechanical interfaces. The simplicity appealed to users, who could just look at the product and know immediately how to use it. Steve Meginniss, the mechanical engineer who worked on the project, had a philosophy, 'The best part is no part.' This philosophy kept us focused on simplifying mechanisms and eliminating unnecessary elements whenever possible."

❷ The form is clean and simple. The bristled head indicates that it is a toothbrush, and the body indicates that it is powered. The differentiated aspect of the toothbrush is not evident upon inspection, though the name on the handle offers a hint. The use of white and light blue give the brush a clinical patina, reinforcing its function as a serious dental hygiene product.

❸ The brush employs two cleaning actions: high-speed scrubbing action, and vibration and cavitation of fluids near the tips of the bristles. The high-speed scrubbing action is achieved by vibrating the bristle head, achieving the equivalent of 31,000 strokes per minute. By contrast, an enthusiastic brusher can achieve 300 strokes per minute with a standard brush, and approximately 4,000 strokes per minute with an electric toothbrush. An effect of this rapid vibration is the agitation of fluids around the bristles, creating waves of pressure, tiny bubbles, and shear forces in the fluids that work to remove bacteria and plaque up to 4 millimeters beyond where the bristles actually make contact. It is this secondary action that makes the Sonicare a revolutionary innovation. Further, the vibrating action is achieved almost entirely through solid-state electronics—there is only one moving part in the toothbrush, making it exceptionally reliable.

❹ There is only one button: it is large, centered in the body, contrasting in color, slightly convex, and rubbery in texture. Later models also indicated that the toothbrush was charging through an indicator light in the button. It is hard to imagine what one could do to make the brush more usable or to make the button better invite pressing.

❺ The body of the brush and recharging base are completely sealed, waterproofing the electronics, minimizing the accumulation of bacteria, and facilitating cleaning. The brush sits upright in the base, enabling it to drain dry while also preventing contact with surfaces. Recharging occurs without any visible electrical contacts, which makes the process seamless and magical. A fully charged toothbrush can last an amazing two weeks without recharging.

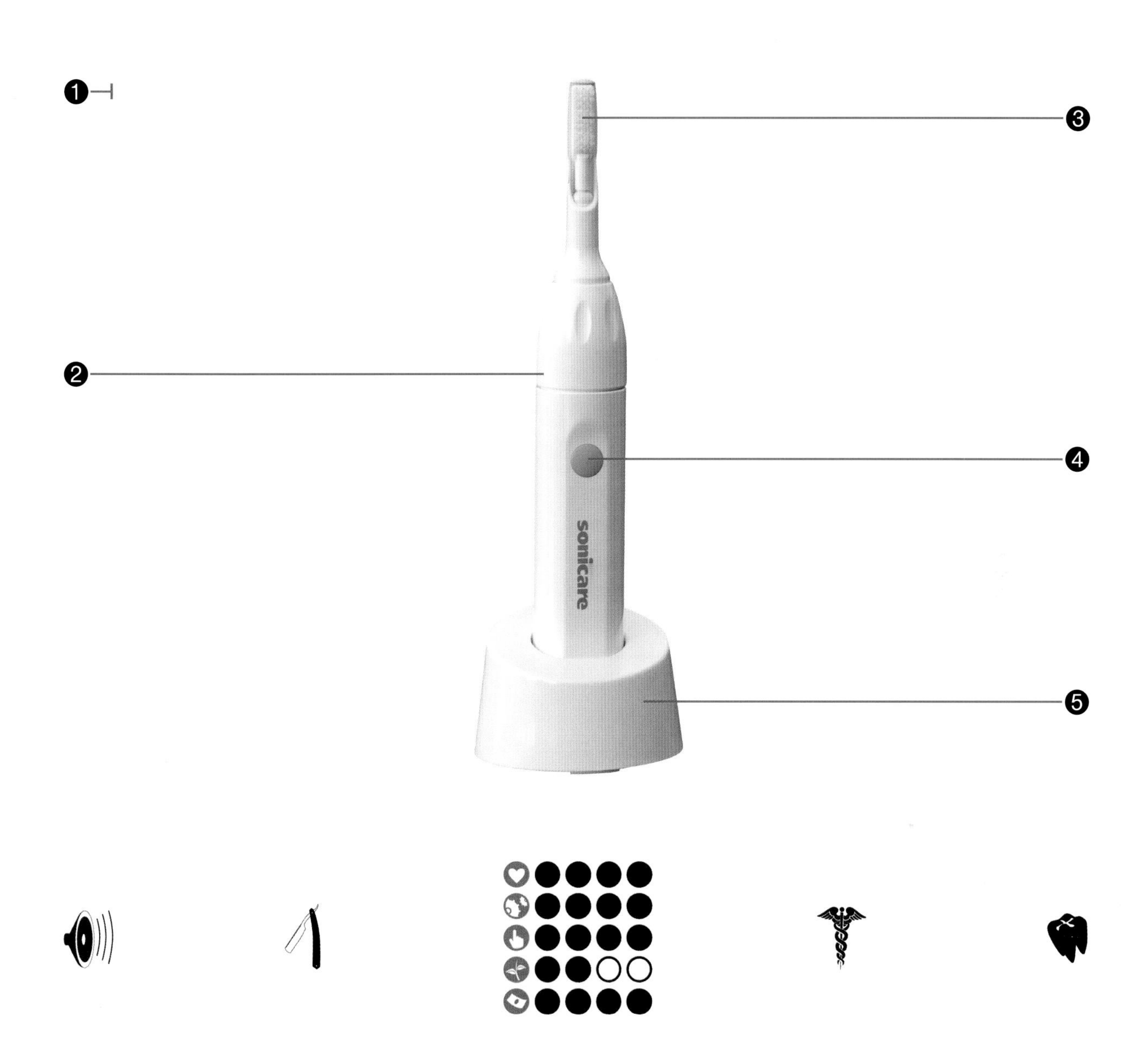
1
2
3
4
5
sonicare

Stand Mixer

Egmont Arens for KitchenAid, 1937

❶ In the early 1930s, KitchenAid commissioned industrial designer Egmont Arens to design a series of new low-cost mixers that would be within the means of the common American household. Arens had developed a reputation for designing for the full range of sensory experiences — an approach he called "humaneering" — making products relaxing to the eyes, pleasing to the touch, muffling noises that jarred the ears, eliminating offensive odors, and so on. This humaneering approach transformed the previously functional but aesthetically unrefined stand mixer into the now classic Model K. Professional cooks and bakers wanted the Model K because of its flexibility and quality. Amateur cooks and bakers just wanted the Model K to be seen in their kitchens — more kitchen art and status symbol than functional mixer. Whether for form or function, you don't mess with a good thing: the Model K Stand Mixer has remained fundamentally unchanged for over seventy years. Egmont Arens comments: "Many inventions such as the locomotive and airplane have slowly developed through a long evolution of added improvements. What happens, however, is that 'added improvements' are often hung on the machine the way added ornaments are sometimes hung on a Christmas tree — as an afterthought. The result is usually a disorganized jumble of wheels, pistons, and cogs which stick out all over the place and which give the machine the appearance of an 'explosion drawing' in which everything seems to fly apart ... Whatever you may be — mechanic, inventor, or engineer — if you create a product which you hope to sell, and want to insure its acceptance to the public, make sure the various parts are organized into a trim, sleek, streamlined shape — for in addition to lowering wind-resistance, streamlining also lowers eye-resistance."

❷ The Model K looks like a Loewy locomotive — stout and streamlined, with a large linear throttle on the side — and is just about as heavy. The mass makes the appliance effectively nonportable, but it also prevents counter creep when beating. Controls are large, intuitive, and well marked. The inner curve of the base and neck precisely follow the curve of the mixing bowl, integrating the forms and giving the mixer a silhouette that is unique and so iconic that it received a trademark from the U.S. Patent and Trademark office. The 1937 Model K came in white only. Colors were introduced in 1955 — Petal Pink, Sunny Yellow, Island Green, and Satin Chrome — further reinforcing the appliance's aesthetic function as a kitchen ornament to be displayed on the counter versus hidden away in a cabinet.

❸ The mixer is equipped with an attachment hub on the front. The hub enables the mixer to redirect its considerable torque to attachments through its nose, converting it into everything from an ice cream maker to a pasta roller. Inserting and removing attachments is a simple affair, requiring the loosening and tightening of a large attachment knob to release and secure the attachments. Every KitchenAid mixer since the Model K has allowed for fully interchangeable attachments, an amazing example of standardization and cross-generational compatibility.

❹ The attachment hub also provides a natural affordance to tilt the head back, which is needed to exchange beaters and remove the bowl. Once activated the beater rotates in planetary fashion — that is, the beater rotates around its own axis while also orbiting the bowl like the earth orbits the sun. Planetary action produces greater contact between the beater and the ingredients, resulting in more efficient and thorough mixing.

1

2
3
4
KitchenAid
0 1 2 4 6 8 10
ARTISAN

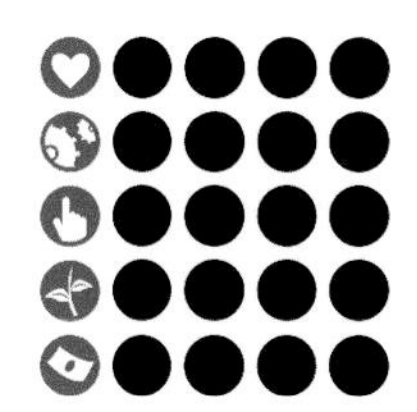

StarTAC Mobile Phone

Motorola, 1996

❶ The StarTAC, TAC for "total area coverage," is the progenitor of the revolutionary RAZR. The phone boasted a number of innovations — smallest and lightest phone of the time, first clamshell design, first mobile phone to have a vibrate option — and, like the RAZR, was frequently seen in the hands of the affluent and celebrity elite. The StarTAC was the flagship product for Motorola in the late 1990s, and the company believed the product would herald a new era of growth and success in wireless technology. Robert Weisshappel, former head of the Cellular Phone Division, comments: "The goal of Motorola has always been to take technology to the next level. Hence, we develop lightweight phones with longer talk times and new capabilities never before been seen, models that directly respond to customer needs ... We have a new portable product that we believe consumers will love. Motorola has developed several new capabilities for it, many of which were previously thought to be impossible, and made them a reality. Just as with the original MicroTAC, we are confident that the StarTAC phone will set the standard for the future of wireless technology."

❷ The phone is the 1990s version of a vintage television series *Star Trek* communicator. It was billed as the first "wearable" phone due to its diminutive size — a size not much larger than contemporary models and nearing the minimum physical envelope of a phone that can be used in a traditional fashion. With the exception of its fragile extendable antenna, the StarTAC was known for its durability, with armored circuit boards and a surprising degree of water resistance. Prices ranged from $1,500 to $2,000, limiting the phone's reach but ensuring its social function as a status symbol.

❸ Unlike modern clamshell designs, the StarTAC located the display on the lower half with the keypad. Standard talk-time was a mere 60 minutes, though "piggy-backing" a second battery could double this time. Buttons are well spaced and provide good tactile feedback. The LED display provides users clear feedback as well as a dense variety of status indicators that modern users now take for granted. The StarTAC was one of the first portable phones to come with a headset, a feature destined to demonstrate the relatively poor multitasking capability of humans — that is, "hands-free" operation does not equal "brain-free" operation.

❹ In 1994, all portable phones were analog, and Motorola owned 60 percent of the market. However, it was clear to everyone in the industry — including many at Motorola — that the world was rapidly moving to adopt digital protocols, which provided superior transmission quality, additional service features, and more efficient use of the radio spectrum. Robert Weisshappel had built the analog business at Motorola. He was convinced that consumers cared more about form factor than the service benefits associated with going digital, and therefore focused Motorola's resources on developing a sleek, feature-rich StarTAC versus a digital technology base. The move delayed Motorola's digital development by years. Competitors exploited the delay and, in a move that would make King Gillette proud, negotiated contracts with network providers to begin giving away their phones in exchange for subscriptions. Motorola, with its high-priced analog StarTAC, lost half of its market share by 1998. As of 2008, Motorola's share of the cellular phone market is below 10 percent. Lessons: 1. Executives, like generals, are typically inclined to fight the last war; and 2. Products cannot succeed by form factor alone.

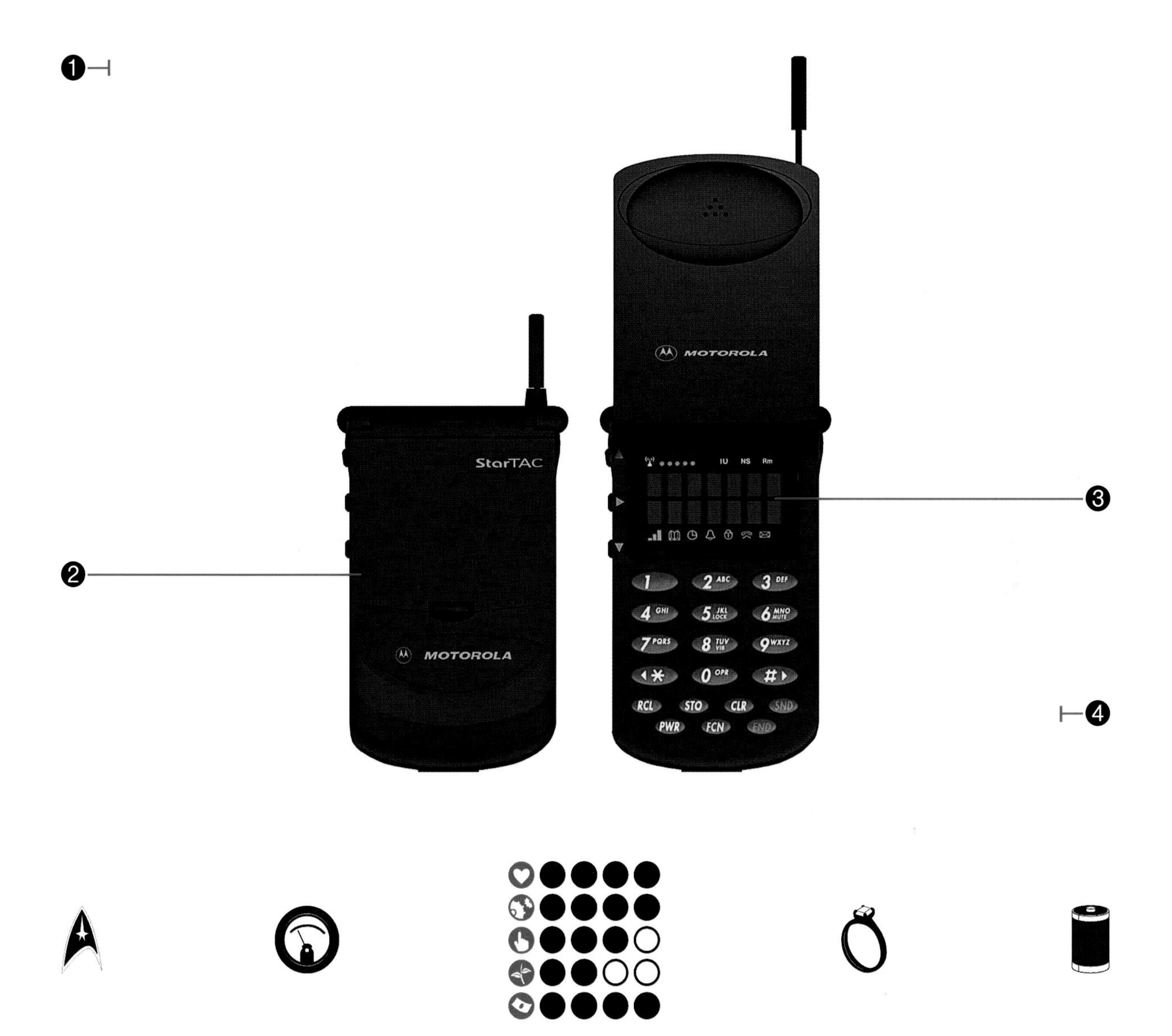
StarTAC
MOTOROLA
MOTOROLA
IU NS Rm
RCL STO CLR SND
PWR FCN END

SX-70 Camera

Edwin Land and Henry Dreyfuss & Associates for Polaroid, 1972

❶ The goal of the SX-70 was to actualize Alfred Stieglitz's vision of removing all barriers between the photographer and the subject, freeing the photographer of all burdens except that of the decision to take the picture. This was a nontrivial goal requiring a nontrivial solution, and it took over twenty-five years to realize it. And long before Steve Jobs was wowing audiences with demonstrations of iPods and iPhones, it was Edwin Land who commanded the room, presenting the new Polaroid SX-70 camera. Edwin Land comments during a demonstration to photographers: "We've looked at the image in the reflex viewer. We've touched the button five times, and we have five dry pictures. While those are snapshots, they're hand-held, not [taken on] a tripod, they're just things as you go along. I think you will sense in them a new meaning for casual photography that makes it not casual, that gives each user a feeling of personal identification with the world around them in the way that photography has always hoped to do, and which it may have done perhaps for many of the people in this room, but which it has not done for the great mass of people who just move on snapping."

❷ Perhaps the most magical product ever invented. The SX-70 required so many innovations in so many different areas working together in a coordinated fashion that is a wonder it was ever invented at all. In its collapsed state, the leather and brushed metal case has but one affordance: grooves on the sides of the viewfinder cap for pulling. A light tug and the slab becomes a camera. The only thing more magical than this transformation is the self-processing film, which produces self-contained takeaway images in just minutes.

❸ The SX-70 set a usability bar that has rarely been matched, especially given its complex functionality. Open the unit by pulling upward. Close the unit by pressing a release arm and pushing downward. To take a picture, plug in a film pack (which also contains the battery), point, focus, and press the red shutter button. The picture auto-ejects and automatically develops. The focus wheel enables the user to focus between ten inches and infinity. Exposure is automatically increased to compensate for faint light. In very dim conditions, flashbulb units can be plugged into a slot above the lens, and are automatically activated when the shutter button is pressed.

❹ The self-processing film is a feat of chemistry and engineering. Each picture is comprised of multiple layers of color sensitive dyes and processing chemicals sandwiched between a transparent cover and black backing, all sealed within a white paper frame. The experience of taking a picture and having a physical photograph moments later is profound even by modern standards. In fact, a significant cost of digital camera technology is the loss of the ability to have an immediate and detachable photographic record. In this way, Polaroid film was as much a social medium as a photographic medium. Pictures could be handed around, discussed, signed, and given away as gifts. Gratification was instant, tangible. While the benefits of digital camera technology have been great, so too has been the social cost resulting from the loss of this capability. As digital camera manufacturers seek ways to advance their market position and forestall commoditization, they need only look to the past to find a high-value, tried-and-true differentiator just waiting to be reinvented.

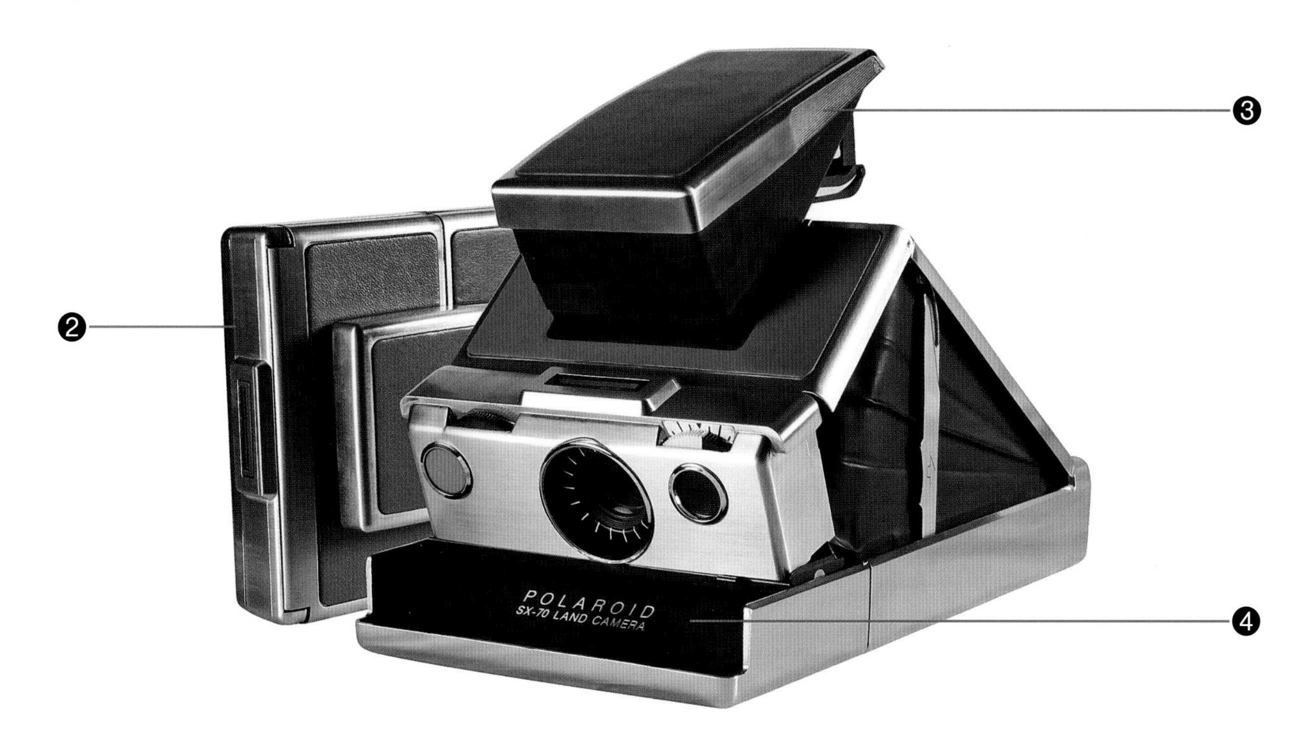

2
3
4
POLAROID
SX-70 LAND CAMERA

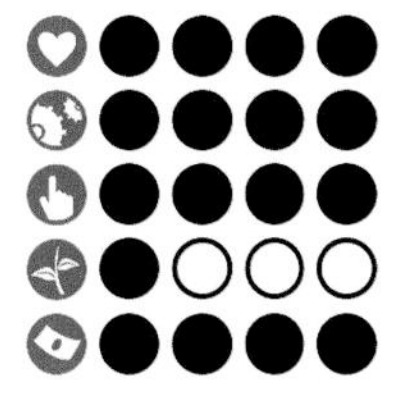

THUMP 2 Sunglasses

Oakley, 2006

❶ William Congreve wrote, "Music hath charms to soothe a savage breast...," but little did he know that music also hath charms to improve training performance. The designers at Oakley did know it, however, and this fact was the driving rationale behind the THUMP sunglasses. Colin Smith, tech guru at Oakley, comments: "Athletes love to train to music. I believe research indicates that performance increases by about fifteen percent if you train to music. So we started to look at the music players that athletes were wearing. They all had what we called the 'ball and chain' problem: a headset attached to a player by a long chord. This is obviously not the ideal situation for an athlete, and it can even be dangerous. For example, if you are a cyclist, you don't want a dangling chord to get caught in the handlebars. Concurrent to all this, we had been doing R&D integrating Bluetooth into our glasses, but Bluetooth was premature for the American market at that time and the project had stalled. Then, at an electronics show in early 2004, we saw a really small MP3 player. It occurred to us that we could take out the Bluetooth module we had been working on, and essentially plug in the MP3 player. Athletes could have the benefits of music without the hazards and inconveniences of the 'ball and chain.' That's how the THUMP was born."

❷ The sunglasses are large and blocky, appearing more like slender goggles than large sunglasses. The aesthetic will be a bit strong for some, but Oakley design has never been for the understated. The glasses are heavy—no doubt owing to the 130-plus electronic components embedded in the frame—but acclimation to the weight occurs quickly.

❸ Integrating disparate products into a single multifunctional product is always dangerous business—either 1 + 1 = 3, or 1 + 1 = 0, it is rarely anything in between. For outdoor athletes who typically wear sunglasses and listen to music—think joggers, volleyball, cyclists, and so on—the THUMP 2 is clearly an example of the former, with excellent optics, rich sound, and no wires. Those not regularly engaged in these kinds of outdoor activities, however, will generally fail to appreciate the value of the integration.

❹ The shape and material of the frame combine to literally stick to the head. The optics as one would expect are excellent, though the thick frames do occlude peripheral vision. The controls are located on top of the arms: volume on the left arm, and playback and power buttons on the right arm. The buttons are easily accessed, and basic functions are intuitive. Advanced functions are available, but are essentially unusable due to their complexity.

❺ The ear buds can be adjusted with six degrees of freedom, accommodating a wide range of variability in head-ear position and shape. Additionally, the buds hold their position, so they can be left to hover just above the ear in case the user needs to hear environmental sounds. Sound quality is excellent, but the speakers are either too strong or poorly insulated from the back and sides—music leaks, even when the buds are in the ears, and anyone nearby will be sharing the musical experience. It should be noted that Oakley likely sees this sound leakage as a positive, not a negative. The THUMP 2 is not just designed for athletes, but also for the same demographic that pimps-out their cars with hyperpowered stereo systems—it is not about listening to music, it is about peacock feathers (i.e., attracting attention).

1

3
2
4
5

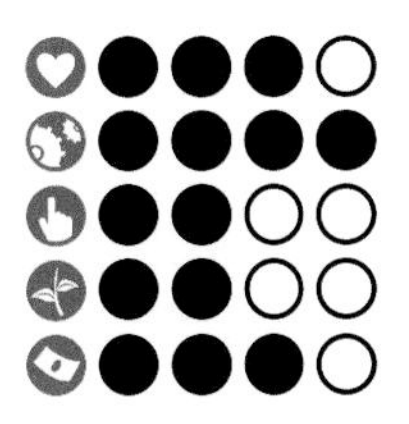

Titanium Series Padlock

Design Continuum for Master Lock, 2002

1. Prior to the Titanium Series, the Master Lock padlock had not been significantly redesigned for more than fifty years. Copied and commoditized by competitors, Master Lock decided it was time to create a padlock design for the twenty-first century, one that could be easily adapted to meet the varying needs and aesthetic tastes of consumers. John Heppner, Master Lock president and CEO, comments: "Padlocks have customarily been approached from a utilitarian standpoint. If you make the assumption that one size fits all, you can achieve efficiencies in manufacturing. But that is a very cost-focused path. I think we've proven that consumers will pay for design. They'll pay for innovation, and they'll pay for improved functionality, as long as it's directed to meet their specific needs."

2. The ovoid form of the padlock is at once aesthetic and formidable in appearance. It fits comfortably in the hand, much like a cell phone, but with the impressive heft of a large, dense rock. The internal mechanism is made of titanium-reinforced steel, the body of stainless steel, and the shroud of injection-molded ABS material. The shackle at the top, the only visible mechanism, suffers minimal exposure and is protected by a reinforced collar. The collar hinders access to bolt cutters and similar devices, but it also hinders flexibility of what you can lock with it—this lock is far less flexible in this regard than its predecessor. The ABS shroud adorns the stainless-steel body like a dress, purposefully exposing the shoulders and arms, leaving no doubt that the hardened body of the lock extends underneath.

3. The internal locking mechanism mimics a bank vault, making use of a planetary gear system that opens and closes a rotary shackle with the turn of the key. This approach effectively resists prying and concussion attacks and eliminates the clumsiness of conventional push-closed padlocks. The only exposed element is a portion of the rotary shackle at the top of the lock, making it the clear target for anyone seeking to defeat the lock—an affordance somewhat mollified by a metal-stamped warning indicating that the shackle has been hardened.

4. The ABS shroud is shaped to sit evenly within the metal shroud, and continues the lines of the shackle at the top. The material minimizes damage caused by the lock (e.g., scratches and dents) and provides it additional all-weather protection. It also enables the manufacturer to inexpensively offer the lock in multiple colors, greatly broadening the potential market for the product.

5. The logo is displayed on the front of the lock just above the keyhole. Given the compelling aesthetic of the padlock, it is a prime location for promoting brand. Unfortunately, rendering the logo with paint is inconsistent with the forged appearance of the other elements and betrays the tactile, three-dimensional language of the form. The result is a reduction in the perceived value of both the lock and the brand—better to have embossed the logo in the shroud.

6. The keyhole faces front (as opposed to down), like a door lock, for improved accessibility. A sliding cover prevents moisture and dirt from entering the keyhole and restores the visual continuity of the shroud when closed. A key-retaining feature smartly prevents key removal unless the lock is fully closed and engaged.

1
HARDENED
3
4
Master
5
2
6

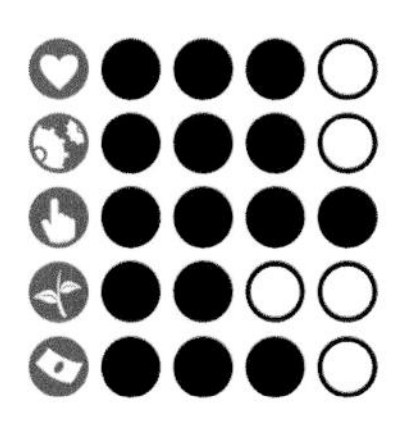

Tizio Table Lamp

Richard Sapper, 1971

❶ "Everyday problems" typically breed "everyday solutions"—the innovation in these cases generally has more to do with recognizing and articulating an unacknowledged problem, than designing and developing a clever solution. Everyday solutions evoke an immediate, "Why didn't I think of that?" response, because the path from problem to solution is relatively straightforward. On rare occasions, however, an everyday problem is solved in such a novel and nonobvious way that it leaves everyone flummoxed. The question asked isn't, "Why didn't I think of that?" rather, "How did anyone think of that?" Such is the case with the Tizio lamp. Richard Sapper comments: "I wanted a work lamp that could be moved by a soft touch of the finger and which would never fall over on my table because of worn-out joints. The usual construction with parallel arms and balancing springs could not meet these requirements because you either have to clamp them onto the table—which is often unpractical—or you have to make do with a limited range of movement—which is just unpractical to me. Because due to my innate inability to keep my desk tidy, I don't have any space for a lamp anywhere near me. The most comprehensive solution to these problems seemed to be a lamp with a structure that is kept in balance all the time by the counter-weights. Then the friction in the joints can be kept to such a minimum that it just balances out the production tolerance in the weights. In order to be effective, any disturbance of the balance system has to be avoided."

❷ The lamp uses a halogen bulb, an innovation in lamps seeded by Sapper's experience in the automotive industry. The low-voltage requirements made it possible to use the arms as conductors, eliminating the need for wires to run power from the base to the bulb. The resulting aesthetic is clean and minimalist, bordering on sculpture. The head shroud, in conjunction with an interior reflective element, focuses light and minimizes glare. The cost of the halogen is heat. Current models cover the bulb with glass to prevent contact, and the adjustment rod doubles as a spacer preventing the head from getting too close to a surface. The problem with using the adjustment rod in this way is the shadow it casts, an unfortunate defect to an otherwise remarkable design.

❸ The Tizio is an expression of pure functionalism, though with an organic, insect-like feel—like a mix between a tower crane and praying mantis. Like a tower crane, its position is kept in a state of equilibrium across all positions by configuring the arm lengths, axis positions, and counterweights such that the sum of the moments always equals zero. Pushing or pulling the adjustment rod creates an imbalance in the system, enabling the head of the lamp to move with minimal effort. Once the adjustment is complete, the lamp resumes an equilibrium state and preserves the position. This approach eliminates the need for tightening knobs and fasteners to hold position.

❹ The base is round and heavy. It houses the transformer and, like the majority of the lamp, is black. The power switch is easily found, in the top front position and high contrast red. The base permits the lamp to rotate a full 360 degrees.

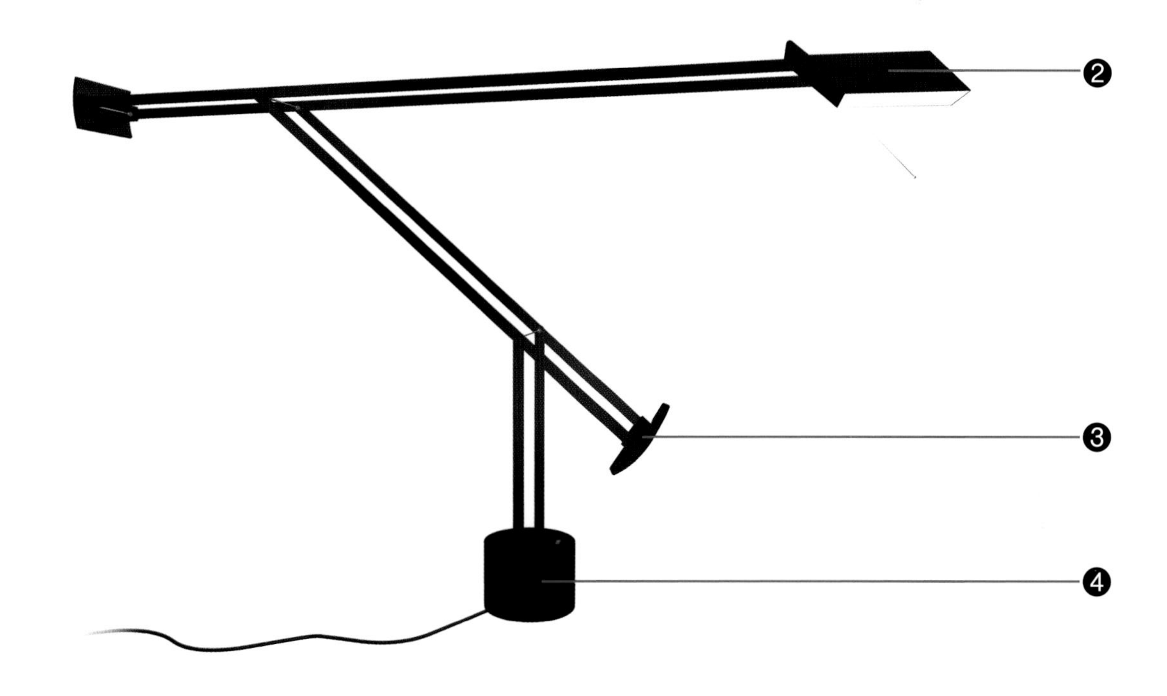

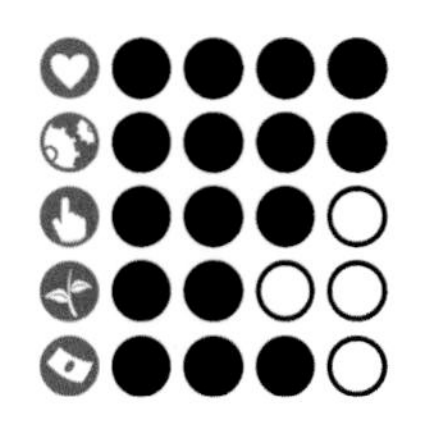

TMX Tickle Me Elmo

Paul Paulsen, Ron Dubren, Greg Hyman, and Sesame Street for Fisher-Price, 2006

❶ Laughter resulting from tickling is not like ordinary laughter—the outward behavior is the same, but the internal experience is very different. Tickling is likely an evolved means of asserting dominance without causing harm, and the laughter that results is a reflexive response that has more to do with indicating submission than enjoyment. With this is mind, it is not so surprising to learn that the inspiration for Tickle Me Elmo came from schoolyard social dynamics. Ron Dubren comments: "The brainstorm for Tickle Me Elmo occurred when I saw a couple of preschoolers having fun tickling one another. It brought back memories of fun tickling jags from my own youth, and, as a toy inventor, I thought it might make a great feature for a plush toy. I pitched the idea to Greg Hyman, an engineer whom I knew already had experience in the emerging sound chip area back in 1994. Together, we developed a working prototype, known then as 'Tickles the Chimp.' The heart of the toy was its laughter and joyfulness, and its soul the emotional resonance we all have to the tickling experience. Essential in the initial conception was the idea of the buildup of laughter, and, eventually, to the payoff of Elmo laughing so hard, he starts to shake."

❷ Elmo's baby face features and high-pitched voice gives him an appealing, non-threatening aesthetic and persona. The tactile aspects of the toy are lacking—he feels like a robot wearing an unpadded Elmo suit. Tickling Elmo, too, is a bit contrived, more like pressing around for a button than tickling. Despite this, Elmo's laughter is infectious and his encouragement for more tickling keeps children playfully engaged, often to the consternation of their parents.

❸ Like all Muppets, Elmo's facial features follow what Muppeteers refer to as the "magic triangle." The magic triangle refers to the relationship between eyes and nose of the Muppet, configured so that the eyes appear to look inward toward the nose, slightly cross-eyed, with a point of focus. The effect brings the face to life and makes the Muppet look like it is actually looking at whoever or whatever is in front of it.

❹ "Laugh and the whole world laughs with you" is more than a saying. It is a scientific fact. Hearing the laughter of others activates regions of the brain that predisposes listeners to laugh along, which is the basis behind laugh tracks in situation comedies. It is this response that the toy exploits, and its effectiveness is a testament to its realistic escalation of laughter supported by a synchronized physical display, which includes bending over, rolling on the ground, kicking feet, and slapping the leg and floor with his hand.

❺ Elmo has three behavioral cycles, each activated by "tickling." The intensity and duration of the laughter and movement doubles with each cycle. Initially, Elmo is passive, and when tickled responds, "You tickled Elmo." For the second cycle, he asks for more, saying, "Again, again." For the third cycle, Elmo has almost had enough, saying, "Give Elmo a break, please." The dialogue is well scripted and the performance flawlessly executed, with Elmo recovering to a standing position at the end of each cycle. Adults will find the interactions briefly amusing, and then grow bored. Young children, however, tend to find Elmo more engaging for longer periods of time—not because he makes them laugh, but because he is something that they can control.

1
2
3
4
5

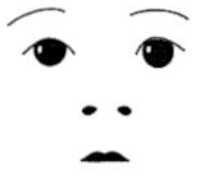

TOUCH! Tableware

Barbara Schmidt, Cornelia Müller, Mirjami Rissanen, and speziell produktgestaltung for Kahla, 2004

❶ It is no small achievement to innovate a product that has not fundamentally changed for more than thousand years, but fusing porcelain with velvet flocking has done just that. The TOUCH! series of porcelain tableware, a name that is cleverly directive and descriptive, demonstrates the role that basic research and technological developments play in product innovation, yielding the all-too-common "solution in search of a problem," Jens Pohlmann, cofounder and designer of speziell produktgestaltung, comments: "To begin with, we were looking for a warming covering for a metal bench, but then we suddenly thought, why not give other products a fluffy surface?"

❷ The TOUCH! tableware is not about porcelain, coffee, tea, or even the development and application of innovative materials, it is about user experience. Given its relative fragility and high price, those who drink their coffee and tea from fine porcelain containers do so for a reason—tradition, taste, social ritual, affluent sensibilities. Introducing change to products that have become integral to the fabric of these kinds of well-established use systems is rife with opportunity and peril. Align a product design with the key vectors of its use system, and, as with Starbucks coffee, the effects can be synergistic, enabling a premium price for a commodity product. Misalign a product design with the key vectors of its use system, and, as with New Coke, the effects can be disastrous. The TOUCH! innovation solves many of the functional problems associated with porcelain tableware. The key questions regard use system conflicts: Will velvet coatings be perceived to be cheap and in poor taste, as many perceive velvet paintings? Will the fuzzy texture and bright colors be perceived to be too frivolous for such serious matters as coffee and tea drinking?

❸ For many traditionalists, subtraction of the telltale tactile and auditory porcelain cues will be anathema to the drinking experience. However, based on all other objective criteria, the fuzzy flocked coating adds far more than it takes away. The hard binary feel of porcelain—either very hot or very cold—has been replaced with a soft and more temperature-neutral material that reduces heat transfer to the hand. Tactile feedback evokes descriptions like "cozy" and "friendly," aligning nicely with friendly meetings over coffee and tea. The fuzzy sensation will be too playful for some, an aspect unfortunately reinforced by unrestrained color choices, but the material goes a long way to redefine the meaning of "fine porcelain." The harsh high-pitched "tink" of porcelain-on-porcelain is softened, so too is dissonance in the experience. The material enables a wide variety of textured colors for the porcelain, and the current palette exploits this flexibility a bit too aggressively. Better to have aligned colors with the tactile-emotional aspects of the design—warm, soft, neutral—than to have drawn randomly from a box of crayons. The result is incoherence in the design and a reduction in the perceived value of this high-priced product, making it more novelty than novel. The shape and location of the material directs users where to and where not to hold and place the tableware. The direction is subtle, welcoming, like a sidewalk in a park—you know you are free to leave the path, but generally don't. The material is surprisingly durable and well affixed, surviving numerous washes and attempts to separate it from the porcelain.

1
2
3

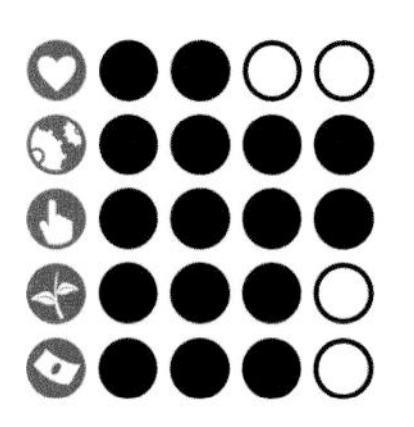

Transistor Radio TR-610

Sony, 1958

❶ When Akio Morita introduced the first Sony transistor radio to retailers in the United States, they incredulously asked him, "Why are you making such a tiny radio? Everybody in America wants big radios. We have big houses, plenty of room. Who needs these tiny things?" Undaunted, Morita pressed the argument that small, portable radios would allow users to take their music with them and enable listening to the kind of music they wanted—the portable radio would individualize music. It turns out Morita was right, and consumers bought the radios in droves. The more portable, the better. The trend toward the small and compact clearly favored Japanese culture at that time (and arguably to this day), a culture preoccupied by the small, light, and compact—an attribute known to the Japanese as *keihaku tansho*, literally "thin and light, convenient and small." Akio Morita comments: "Miniaturization and compactness have always appealed to the Japanese. Our boxes have been made to nest into one another, our fans to fold, our art rolls into neat scrolls."

❷ The TR-610 is the progenitor of both the Walkman and iPod. Its plastic case is contoured, designed to fit the hand, an ergonomic consideration not seen prior to this model. Both the tuner and volume rotary controls are adjustable using the thumb (for the right-handed, that is), with the station frequency displayed center front just below the logo. A large circular speaker shielded by a perforated metallic cover dominates the front, such that users would hold the radio against the side of the head with the speaker against the ear. The TR-610 was available in four colors—red, ivory, black, and green—to appeal to younger consumers who were the predominant buyers of the radios.

❸ Timing, as they say, is everything, and so it is with the transistor radio. To understand the rapid rise and success of the transistor radio, specifically the TR-610, it is necessary to understand the relevant social forces at work in America during the mid-1950s—and they can be essentially summed up in one name: Elvis Presley. In 1954, Elvis propelled rock and roll to the fore in the United States with the number one hit single, "That's All Right." Teenagers loved the new style of music. Adults, not so much. The transistor radio allowed teenagers to listen to what they wanted, wherever they wanted, out of earshot of fussy parents. The success of the TR-610 is the direct result of Sony's response to this social trend: bright colors, a speaker that looks louder than it really is, and an audacious post-modern design. Previous models had incorporated the Sony name into the design—a name derived from the Latin *sonus* and the more colloquial "sonny boy"—but the TR-610 was the first to incorporate the current serif logotype developed by Yasuo Kuroki. The transistor radio became an expression of individuality and independence for teenagers in the 1950s, much like the mobile phones and iPods of today.

❹ Sony emphasized the pocketability of the radio, requiring its sales staff to wear oversized shirt pockets that could contain the radio for demonstration purposes. The hinged metal stand further reinforces the point. The stand rotates out slightly from the base to enable the radio to stand upright for stationary listening, as well as swing to the top, where it becomes a handle for carrying. Many advertisements featured fashionable women with the radio in hand, stand extended, making the radio resemble a tiny purse.

SONY
SIX TRANSISTOR

Tykho AM/FM Radio

Marc Berthier for Lexon, 1997

❶ In his explorations of media, Marshall McLuhan asserted that "the medium is the message," meaning that the medium itself communicates information and ultimately influences the interpretation of the message. There is perhaps no better example of this observation in product design than the Tykho radio. Given its toylike patina, the radio has received a surprising degree of acclaim, including induction into the Museum of Modern Art, featured on the cover of the *Time* magazine issue focused the rebirth of design (the subhead was "Function is out. Form is in"), and recognized by the Pompidou Center in Paris as one of the two objects that best signify the years 1985–2000—presumably because it looks "designed." To what can this success be attributed? The price is relatively high. The functionality is minimal. The performance is mediocre. The aesthetic is, well, different, and to quote Robert Frost, "That has made all the difference."

❷ The composition of elements is simple and organized. The case is basically divided into two quadrants, the left containing the speaker and power button and the right containing the volume and tuning controls. All elements are center-aligned within their respective quadrants, making the already simple form appear even simpler. The case is ABS plastic covered by a silicon rubber shell. Given the seamlessness of the rubber cover in the front, it is disappointing to see the four metallic case screws exposed in the back. Features are bare minimum. There are no programmable channels. No earphone jack. No lighted displays. No handle. Tuning is achieved by turning the antenna, a clever adaptation but visually and functionally disjointed from the other controls.

❸ The rubber gives the radio an aesthetic that is difficult to reconcile. On the one hand, it gives the radio an interesting and original aesthetic that explores the application of injection-molded rubber to achieve a new level of minimalism for electronic devices. On the other hand, the bright colors and the surface homogeneity, coupled with the device's simplicity, make the radio look like a cheap toy. They also, however, make the product look like a physical manifestation of an icon—icon in the graphic design sense of the word—that gives the radio a memory stickiness not found with other products. This is not sufficient to drive mass-market sales, but it is more than sufficient to get the attention of editors and writers of design periodicals (and books). This is not to suggest that there is not some worthy design in the radio aside from its rubber exterior. The buttons are all clearly marked and well grouped. The decision to make the AM/FM controls two independent buttons and the two volume controls one linked button correctly reflects that the former are distinct functions whereas the latter represents two ends of a single continuum—and their layout together resembles a face, which is rarely a bad thing. The adjustable antenna is a clever and original means of tuning. The Tykho also boasts both splash and shock resistance, but it is clear that it is not intended nor equipped to be a serious outdoor or survival-grade radio. So, what kind of market is there for a well-designed, nominally capable, high-priced AM/FM radio? Not much of one. There is, however, a market for inexpensive, functional "modern art" among designerati, which is the population this product clearly targets.

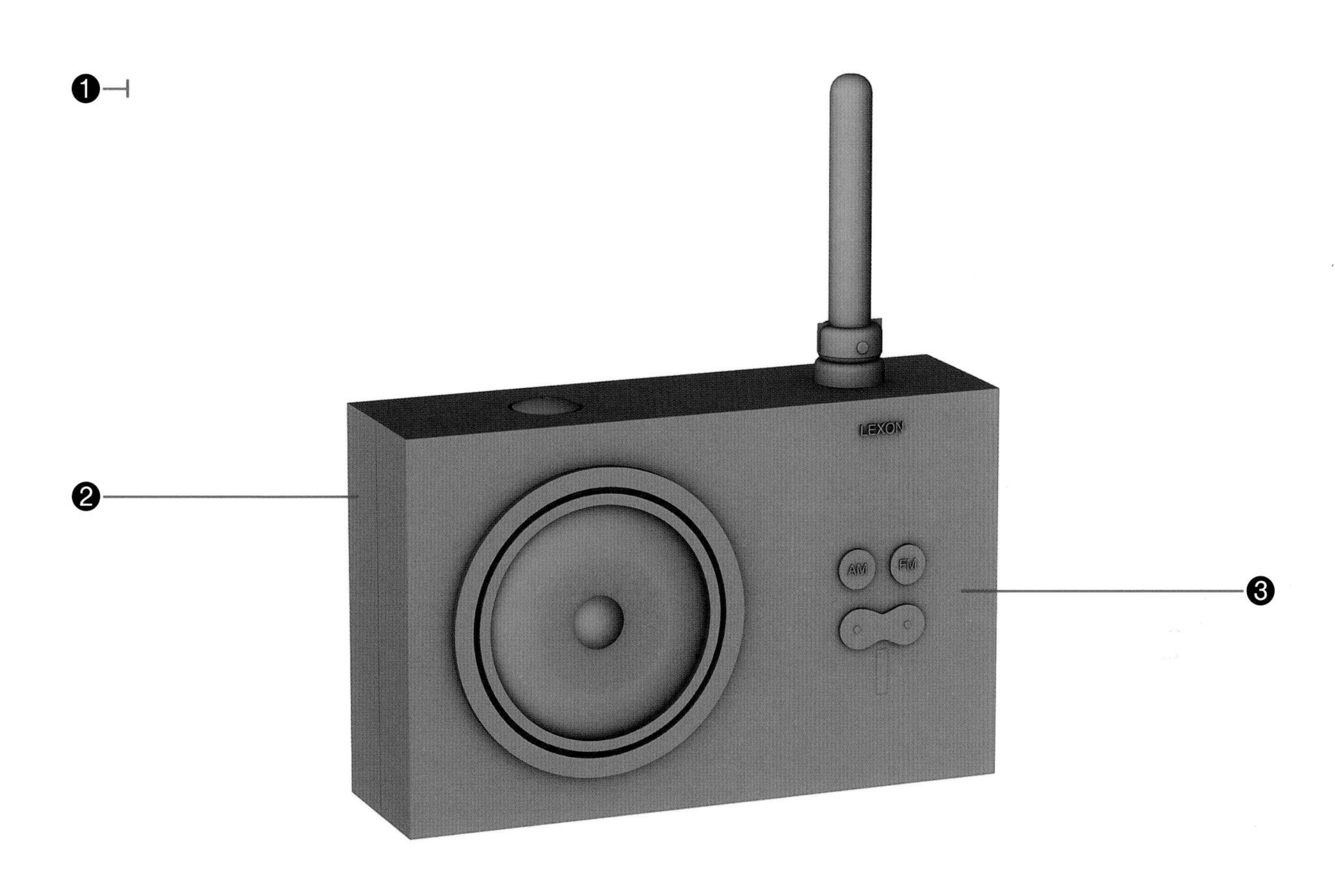
1
2
3
LEXON
AM
FM

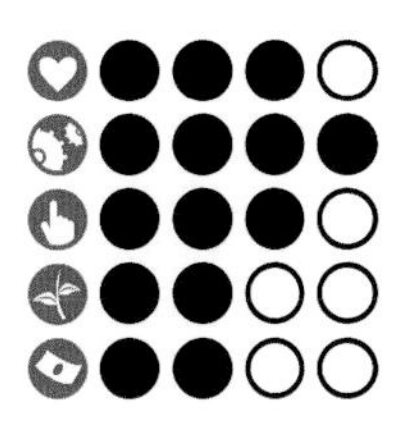

Variable Balans Stool

Peter Opsvik, 1976

❶ Peter Opsvik has been rethinking the way people sit since the early 1970s, challenging virtually every assumption of chair design, both form and function, considered sacrosanct by furniture designers. With the Balans, a variation of the word *balance*, he presses the case that the flaw of traditional chair design is its failure to address the natural urge and ergonomic need to keep moving when seated. Peter Opsvik comments: "If we are able to move, we do. When we stand up, we very rarely ever stay still—we shift our weight around, moving our feet and our arms much more than you ever think. We can walk for hours, but we get tired after only a few minutes if we have to stand still."

❷ The aesthetic is utilitarian, akin to that of a rocking chair sans the back support, and offers little by way of affective appeal. Assuming the primary sitting position—sitting on the upper pad with knees on the lower pads—is surprisingly intuitive, though entry and exit are cumbersome for the newly initiated and the infirm. The stool is designed to promote movement and variation, not just to be "sat in" or "kneeled in," and accordingly compels the user to self-balance and self-support their posture. It is this "active" paradigm of sitting that is its greatest challenge, as it flies against convention of what sitting is understood to be—that is, resting in a supported position and not moving. Accordingly, appreciating the design is as much a process of education as exposure. The user must buy into the basic proposition that "active sitting" is good for you, even if it isn't always comfortable for you.

❸ The primary seating surface is large and well padded, parallel to the upright position of the rocker arms. The surface is flat, allowing the user to shift, spin, and slide in and out, though as stated this maneuver requires practice. The kneepads enable users to achieve steeper leaning positions without sliding off the seat than would otherwise be possible with traditional chairs, making it is easier to retain the natural curvature of the lumbar region.

❹ The two kneepads are appropriately angled, supporting average statures with little or no discomfort, and facilitating changes to leg positions. The kneepads clearly afford having knees and upper shins placed on them, which is one reason many users assume this is the singular manner that one is supposed to sit in the stool. It's not. Rather than trying to promote one Platonic ergonomic posture, the stool aspires to accommodate them all; the premise being that even a highly comfortable position becomes uncomfortable over time. Therefore, users are supposed to actively change their sitting angle and position, sometimes using one versus two kneepads, sometimes using neither. This manner of use is intuitive, but leads many to think that they are working around a design defect versus using the stool as intended.

❺ The rocker arms support two resting positions, leaning forward and sitting upright, and a third balancing position at the fulcrum. The third position is the most interesting and distinctive, underscoring the "balance inspires movement" philosophy of the design. Used properly, the dynamic balancing and rocking motion is not just healthy, but enjoyable.

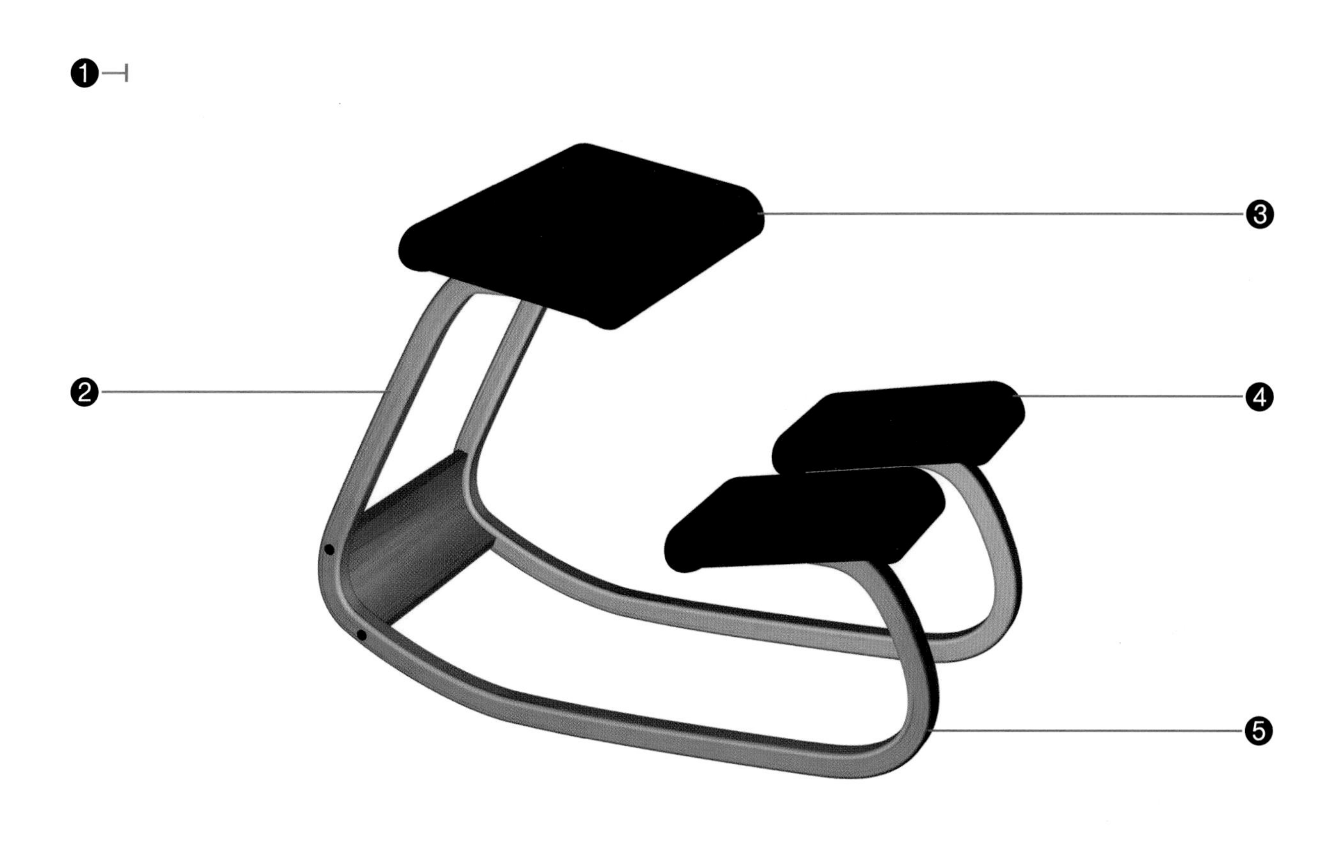
1
2
3
4
5

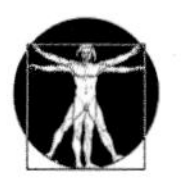

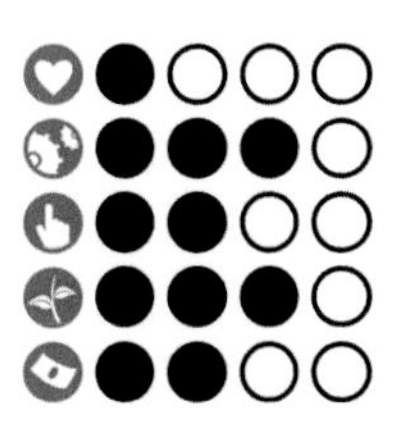

Vespa 150 GS

Corradino D'Ascanio for Piaggio, 1955

❶ To paraphrase Mark Twain, often the greatest barrier to innovation is not what we don't know, but what we do know that just isn't so. Breakthrough innovations often require individuals who possess strong generalist design and engineering knowledge, but who specifically lack experience with a particular product domain. Such generalists are able to look at problems fresh, uncorrupted by past assumptions and conventions, and leverage analogical knowledge from other domains. Such was the case when Enrico Piaggio sought out noted aeronautical engineer Corradino D'Ascanio to develop "a vehicle that will put Italy on two wheels." D'Ascanio comments: "Not knowing motorcycles, I was in the ideal position to create a vehicle without precedents. [Enrico] Piaggio was counting on this. 'Only you can tackle the problem with a wholly new outlook,' he told me. I followed intuitive criteria. I felt that the machine ought to serve those who, like me, had never ridden a motorcycle and hated the machine's lack of maneuverability. I thought it over for a while and on Sunday the basic idea came to me. The most important factor was being able to mount the vehicle comfortably, something that had already been resolved with the ladies' bicycle. So I started out with the concept that is fundamental in the ladies' bicycle. I felt a seated position was more comfortable and rational than having to straddle the frame."

❷ Reinterpret a light aircraft as a wingless two-wheel vehicle, and you get the Vespa. The body is a monocoque versus frame design, a structural configuration generally reserved for aircraft. The "nose" wheel is mounted to a mono-strut and shrouded by a fairing. Wheels are small, giving the vehicle a "cute" quality. The pass-through leg bay makes the vehicle seem approachable, friendly.

❸ In many ways, the Vespa is the progenitor of the Segway; both vehicles are attempts at addressing problems of urban navigation through highly unconventional means. The Vespa's low center of gravity, small wheels, and handlebar design made it highly maneuverable, ideal for urban use. Gearshift, throttle, and front brake are all conveniently located on the handles, keeping the hands on the handlebars and avoiding scuffs to fine Italian shoes. A fairing shrouds the handlebar mono-strut, giving the vehicle an aerodynamic appearance and protecting the rider from wind and road debris.

❹ Unlike motorcycles, the engine of the Vespa is completely concealed. The primary rationale was to prevent the rider from dirtying their hands and trousers, but hiding the engine within the monocoque shell also gave the Vespa a highly differentiated, streamlined aesthetic. The gas cap is located beneath the hinged saddle seat. The two-stroke engine produced 8 horsepower, copious amounts of smoke, and a high-pitched buzzing sound—Piaggio related the sound to that of a wasp, hence the name Vespa, the Latin and Italian word for "wasp"—but the scooter's aerodynamic shape and lightweight made for a zippy ride with a top speed of 60 mph.

❺ The pass-through leg bay is perhaps the greatest innovation in terms of form factor. The Vespa was the first motorized vehicle to specifically consider the needs of women in its design, accommodating skirts and dresses as well as pants. The flat floorboard offered protection from the ground, and did not discriminate against heeled shoes as do the pedals and pegs of bicycles and motorcycles. The large saddle seat not only provided increased comfort over earlier models, it encouraged couples riding together, making the Vespa a popular social form of transport for younger riders.

Walkman TPS-L2

Sony, 1979

❶ Progenitor of the now ubiquitous iPod, the Sony Walkman was the first practical device that enabled users to listen to the exact music they wanted, when they wanted, where they wanted. This user requirement was perhaps first understood by former Sony chairman Masaru Ibuka, who was often seen toting a large tape deck with bulbous headphones so he could listen to his music when he traveled. Akio Morita, founder and honorary chairman of Sony, took note of the extremes his colleague was willing to endure to make his favorite music portable, and directed his staff to make a small experimental cassette player with light, comfortable headphones. The Walkman was born. Akio Morita comments: "The public does not know what is possible, but we do. So instead of doing a lot of market research, we refine our thinking on a product and its use and try to create a market for it by educating and communicating with the public... I do not believe that any amount of market research could have told us that the Sony Walkman would be successful."

❷ Headphones prior to the Walkman were typically the large over-the-ear variety, weighing in excess of 300 grams. The Walkman introduced on-the-ear headphones that weighed a mere fifty grams and produced comparable stereo quality sound. While many might consider the design of the headphones unnecessary relative to the design of the player, it is a testament to the designers that they understood that the size and weight of the headphones had to be visually proportionate to the player for it to be considered portable.

❸ Concerned that the device would be perceived as an anti-social technology, Sony incorporated two headphone jacks, labeled "guys" and "gals" respectively in some markets, to promote collaborative listening. This, of course, raised the concern that users could become detached from their surroundings while listening to music, and so a gimmicky "Hot Line" feature was added, enabling users to mute the music and amplify ambient sounds. Sony soon realized that empowering users to be politely antisocial was actually one of the reasons for the Walkman's success, and both features were dropped in later models.

❹ The rectangular form is functional, but aside from its small size, unspectacular. Attention to detail gives the device a feeling of quality—metal construction, leather carrying case, hinged battery cover—but gains in this regard are undermined by its gray-blue color and excessive labeling. Ulm School, where were you? Independent volume controls for the left and right channels are located below the playback controls on the side. The fast-forward and rewind functions provide high-speed audible feedback, enabling users to quickly and efficiently navigate across songs.

❺ The genius of the design regards aligning the feature set with the user requirements for portable players, and abandoning functions that were unnecessary. Specifically, eliminating the recording capability and loudspeaker was considered unthinkable at the time, and it was strongly resisted by the internal sales team at Sony. Akio Morita responded by declaring that he would resign if the product did not succeed. The greatest design lesson of the Walkman has little to do with the elements of design embodied by the product, and a lot to do with the kind of leadership required to create truly revolutionary products in large organizations.

1
2
3
4
5
SONY
STEREO CASSETTE PLAYER TPS-L2
STEREO
SONY
R

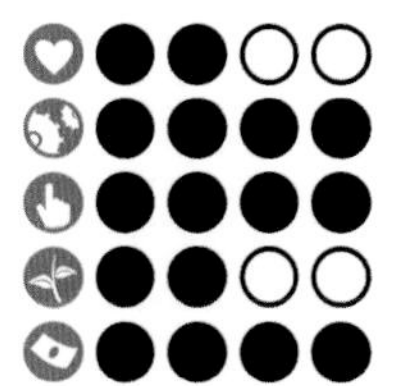

Wiggle Chair

Frank Gehry, 1972

❶ The Wiggle Chair, an homage to Reitveld's Zig-Zag Chair, was developed as part of the Easy Edges series of furniture, the series name derived from its namesake material, Edge Board, a material comprised of glued layers of corrugated cardboard running in alternate directions. The abstract and sculptural series achieved critical success and Gehry received wide acclaim as a furniture designer. Frank Gehry comments: "The nice thing is you can just pick a piece off and throw it away if you don't like it. I had made some chairs earlier, and they were shown at Bloomingdale's. I made them out of paper. They achieved some kind of commercial success and it scared me, so I stopped them, because I wasn't ready to be a successful furniture designer. I still wanted to be an architect. Somehow I thought that was going to end my life, so I stopped them, and I started making chairs that I thought nobody would like, and that's what these are."

❷ The chair is generally considered more art piece than functioning chair, with its cardboard construction, snake-like profile, and architectural heritage. The form is a single length of thick material folded and looped to form a chair. The material is a variant of cardboard, a honeycomb paper sandwiched between two hardboard structural edges. The chair is light and surprisingly strong, flexing slightly when the user sits and leans backward, but never giving the impression of instability. The material has the texture of rough velvet or corduroy, porous and airy, but not conducive to long-term sitting. Aside from the appearance of instability, reinforced with a name like Wiggle, the chair fosters a positive affect. It grabs the attention in a room and invites people to come sit—if only for a moment.

❸ The Easy Edges series of furniture was conceived as a seventeen-piece set of low-cost, mass-market furniture, with individual pieces selling for as little as $15. Furniture was made of commonly available recyclable materials and required simple manufacturing processes to mass-produce—considerations that dominate product design today. However, the Wiggle Chair is unique in the series in that it is the only piece that does not try to mimic the form of traditional furniture; rather, it is a true expression of the material of which it is constructed. While one can perhaps credit Reitveld with the innovation of form, it is the expression of that form where Wiggle transcends the original—the folds in the form tell a clear story about the nature of its material—simple, flexible, inexpensive, sustainable—while still fulfilling the functional requirement of supporting the seat. The chair also possesses a uniquely organic aesthetic, a Gehry trademark and a stark contrast to the Zig-Zag, that resembles a large constricting snake poised to strike. Humans are innately wired to detect snakes and snakelike forms. One wonders to what degree, if any, the Wiggle triggers these innate cognitive programs, ensuring that passers by take notice and pay attention. Could this be one of the reasons so many people know of the Wiggle, and so few know of the other pieces in the set?

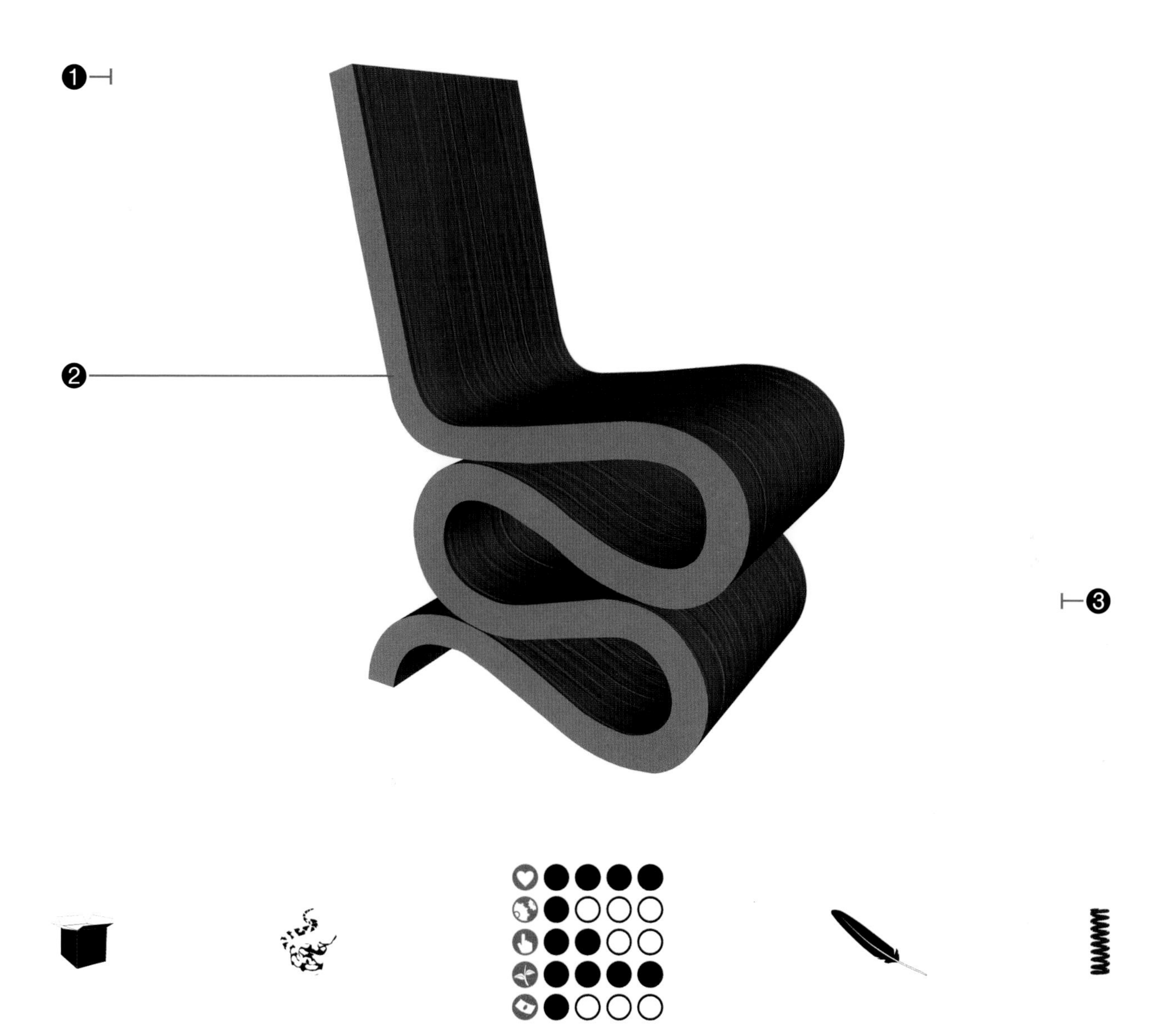

Wii

Nintendo, 2006

❶ What does it look like to have fun? This is in many respects the most basic question for game design, but it was rarely if ever discussed—until, that is, Nintendo introduced the Wii. Should a game system encourage players to sit alone, hunched in dark rooms in front of televisions, intensely gripping controllers and thumbing buttons, or encourage players to stand with friends and family, in lit rooms, casually moving their arms and bodies in what appears to be dancing or calisthenics? Shigeru Miyamoto, general manager of entertainment analysis and development at Nintendo, comments: “I think what’s really important is to think of how the player feels while they’re playing. For instance, with the Tennis games, you don’t necessarily have to do big swinging motions to play it, you can actually make just very simple motions; you could even just tap the controller back and forth on your hand and still execute the actions on the screen. But in fact for most players getting a good swing in and actually playing the game with those sweeping motions is a lot more fun. Simultaneously, something else that we’ve tried to think of is, as we’re creating the games, is does the game look like it’s fun to play when you see someone else playing it? I think that’s very important, this idea of when other people are looking at the player are they being encouraged to actually try and play the game as well.”

❷ The form of the console is unremarkable, and apparently this is by design. The requirements were to develop a form factor that “people would not hate” about the size of a few standing DVDs, such that it could sit beside the television in a crowded entertainment system.

❸ The success of the Wii can be summarized in three words: human-machine interface. Nintendo observed that the game industry was engaged in a kind of arms escalation, where success was defined by the power of the hardware and sophistication of the software, both of which were advancing at staggering rates. The resulting trends: increasing price competition, thinning hardware margins, narrowing user demographic, and geometrically rising game development costs. Accordingly, Nintendo focused their design energy away from the focus of the competition and on the human-machine interface, which had not fundamentally changed since the joysticks of the 1970s. The new interface is based on wireless hand controllers with accelerometers and infrared sensors. The accelerometers and sensors convert arm and wrist motions into signals that are then transmitted to the console, where they are converted to game actions. The controllers, in combination with relatively simple and collaborative games, are intuitive to use and appeal to a much wider audience than traditional joystick-based handsets. The system, however, is very first generation. Slow movements translate well, but rapid movements often yield ambiguous results. The controllers are easily held, but neither look nor feel like objects that want to be held. They possess a pedestrian aesthetic—functional but boring. Basic sports games naturally exploit the physicality of the interface, directly mapping arm and hand motions of the user with arm and hand motions of an on-screen character. More abstract games lacking intuitive physical gestures, however, are far less compelling, though it will no doubt take some time for interaction and game designers to fully understand and exploit the possibilities afforded by this interface. Straps were added (and later strengthened) to prevent inadvertent controller launches.

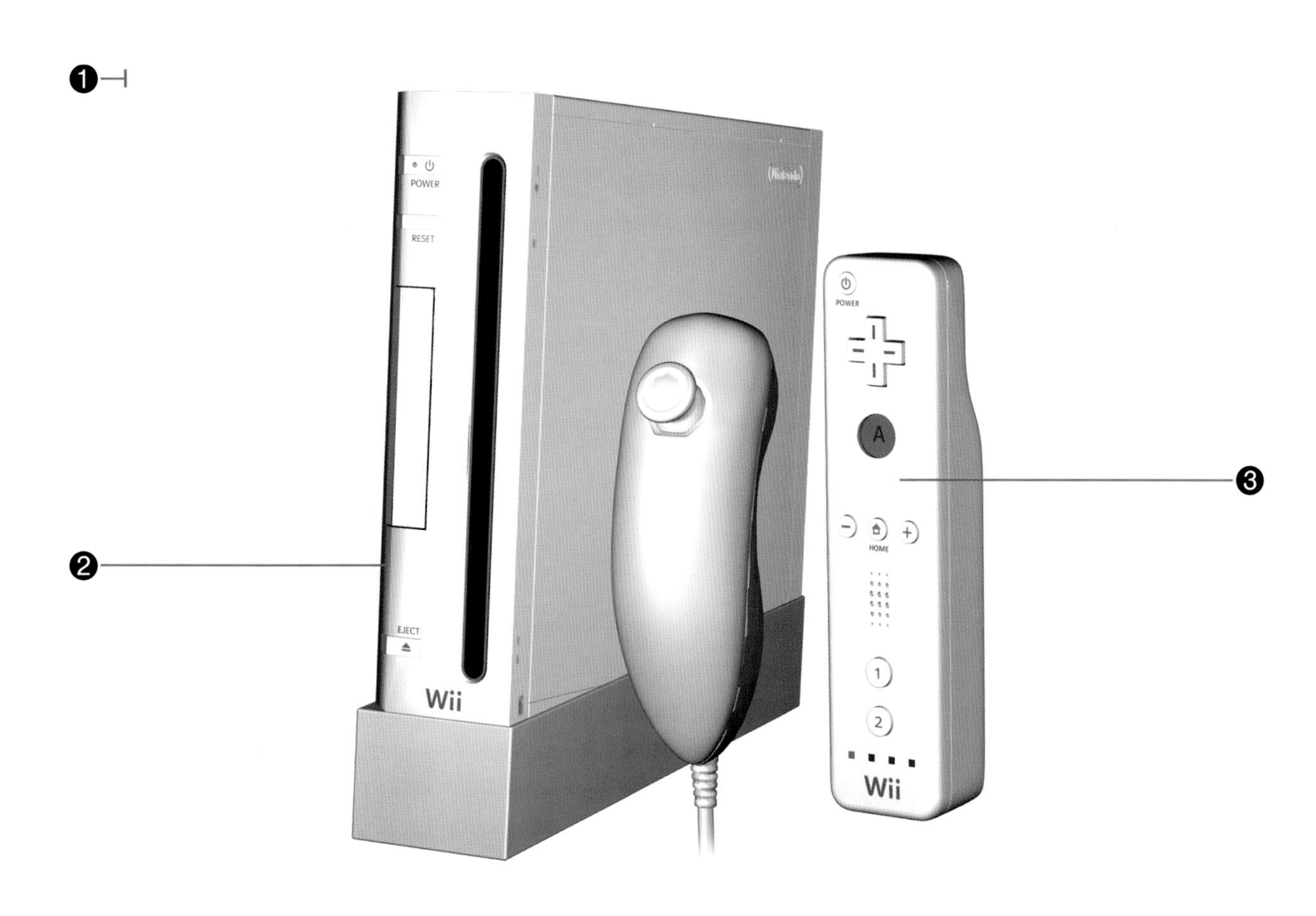
POWER
RESET
EJECT
Wii
POWER
A
HOME
1
2
Wii

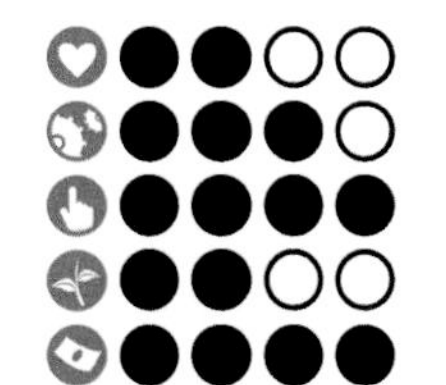

XO-1 Laptop

Fuseproject and One Laptop Per Child Association, 2005

1. Many learning experts believe that effective education is as simple as giving children the tools and opportunities to learn, and then staying out of their way. This approach makes the learner a stakeholder in their education, giving them a greater locus of control and their learning activities a greater sense of purpose. Nicholas Negroponte, founder and chair of OLPC, comments: "Most people think that kids drop out of school in the developing world to go to work and earn money for their family or to work in the fields or to take care of the siblings or something like that. In fact, that happens—but the primary reason is that school is boring. It's not relevant... This is why we make it a requirement that the kids take the laptops home, that they own the laptops. It seems onerous to some countries, but if they don't want the kids to own the laptop, we won't do business with them."

2. In the closed position, the XO looks like a thin lunchbox or medical kit. It is small, approximately half the size of a typical laptop, encased in a ruggedized green and white plastic body. The object is unique and easily recognizable from a distance. A handle and strap holes are built into the case. The back of the monitor has a large XO logo, which comes in a variety of color combinations so that laptop can be somewhat personalized. The small size and bright colors make the object desirable to children and undesirable to adults, a key consideration to minimize theft by adults. The outer case is textured with goose bumps, protecting it from scratches and giving it a playful, tactile signature.

3. The display pivots between three positions: closed (display protected), open and upright (display in traditional computer position), pivot and closed (display in e-book reader position). Wi-Fi antennas, looking like ears when extended, lock the display in the closed positions, protecting the various data ports. The transition is slick, and the e-book reader position is particularly intuitive and useful. Designed for use in outdoor classrooms, the screen is readable in direct sunlight, and allows the user to toggle between color and black-and-white modes to conserve energy. The amount of technology stuffed in this little box is extraordinary—Wi-Fi, mesh networking, microphone, video camera, and so on.

4. The software is the clear weak link in the design. The user interface, named Sugar, is not intuitive, and the applications, many of which are genuinely creative, are unpolished. The usability deficits are serious. No doubt a determined child can achieve superficial use quickly, but actually using the device to collaborate or learn anything meaningful requires considerable investment. The oft-invoked argument that the interface was designed for children and that is why adults find it baffling, is simply not credible. The look and feel is circa 1980s, and the performance is not much better. Proponents assert that the XO's cheap price does not mean cheap quality, but this is not true of the software.

5. The keyboard is covered by a single, rubber membrane, making it water and dust resistant, reminiscent of the Sinclair ZX-80. Language localization is inexpensively achieved by printing different characters on to the membrane. Keys are in standard QWERTY configuration, though far too small and spaced too closely to be used by adults. It is very clearly a child's keyboard. A large, sealed touchpad makes for intuitive cursor control.

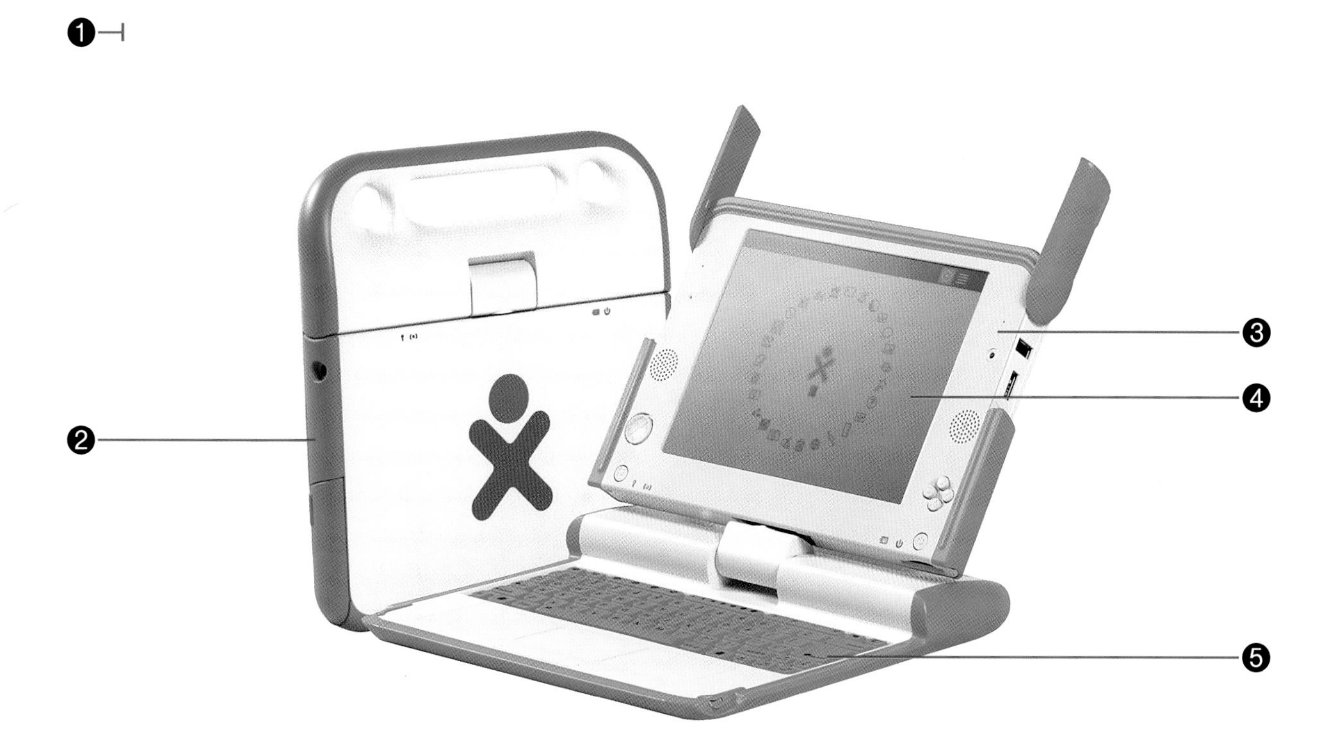
❶
❷
❸
❹
❺

100

Zippo Cigarette Lighter

George Blaisdell, 1932

❶ Great design is sometimes born of necessity, sometimes opportunity, and sometimes both—so it is with the Zippo lighter, named after the then-recently invented zipper for no other reason than he liked the way it sounded. Sarah Dorn, daughter of Blaisdell, comments: "The one thing [my father] did know in the early '30s was that he needed to do something, because those days were tough. There wasn't a lot of money lying about then, but he went to everybody to get the money to launch Zippo Manufacturing Company. Nobody had any faith in it. It seemed like a foolish, harebrained idea. And it was. Imagine: manufacturing and marketing a lighter for $1.95 when that amount of money fed a family. What kept him going? I think whatever it was, it was tinged with desperation. He had to make this work. For him and his family, as they say now, there was no Plan B."

❷ The form would make Walter Gropius proud—simple and modern. The design is a refinement of an Austrian model Blaisdell observed a friend using at a local country club. The Austrian lighter functioned well, but suffered a number of design deficits: utilitarian aesthetic, required two hands to use, thin metal shell that dented easily. Blaisdell acquired sole U.S. rights from the Austrian lighter manufacturer and introduced a number of improvements, including its distinctive rectangular shape, strengthened casing, hinged versus removable top, and a punched metal chimney to protect the flame against wind. The resulting design, comprised of a mere twenty-two components, continues to be popular worldwide and remains virtually unchanged after more than seventy-five years.

❸ One cannot fully understand the design of the Zippo lighter without considering the key role that marketing played in its success, and nothing impacted its success more than its lifetime guarantee. Zippo positioned itself as a premium lighter, and charged an appropriately premium price. To help justify the price point and differentiate the quality of the product from lesser brands, Blaisdell offered the following unconditional lifetime guarantee to buyers: "It works or we fix it free." Damaged or defective lighters sent to Zippo were returned postpaid within forty-eight hours with a note reading, "We thank you for the opportunity of servicing your lighter." The manufacturer claims that in more than seventy-five years, no one has ever spent a cent on the mechanical repair of a Zippo lighter, regardless of the lighter's age or condition. This unequivocal service component combined with the product's brand positioning underscored its differentiation, and ultimately enabled its success. Zippo was also dramatically impacted by World War II, during which time the company ceased consumer manufacture and focused strictly on manufacture for the U.S. military. Their wartime contract enabled Zippo to reach millions of military personnel, greatly expanding its reach and cultivating a loyal postwar consumer base. The widespread use of the lighter in the war—a time of great patriotism, camaraderie, and historic sense of achievement—made it the beneficiary by way of classical conditioning of many of these and related affects. The sight of the lighter, the distinctive clicking sound of its activation, the feel of it in the hand, all associative memory triggers for a generation at war.

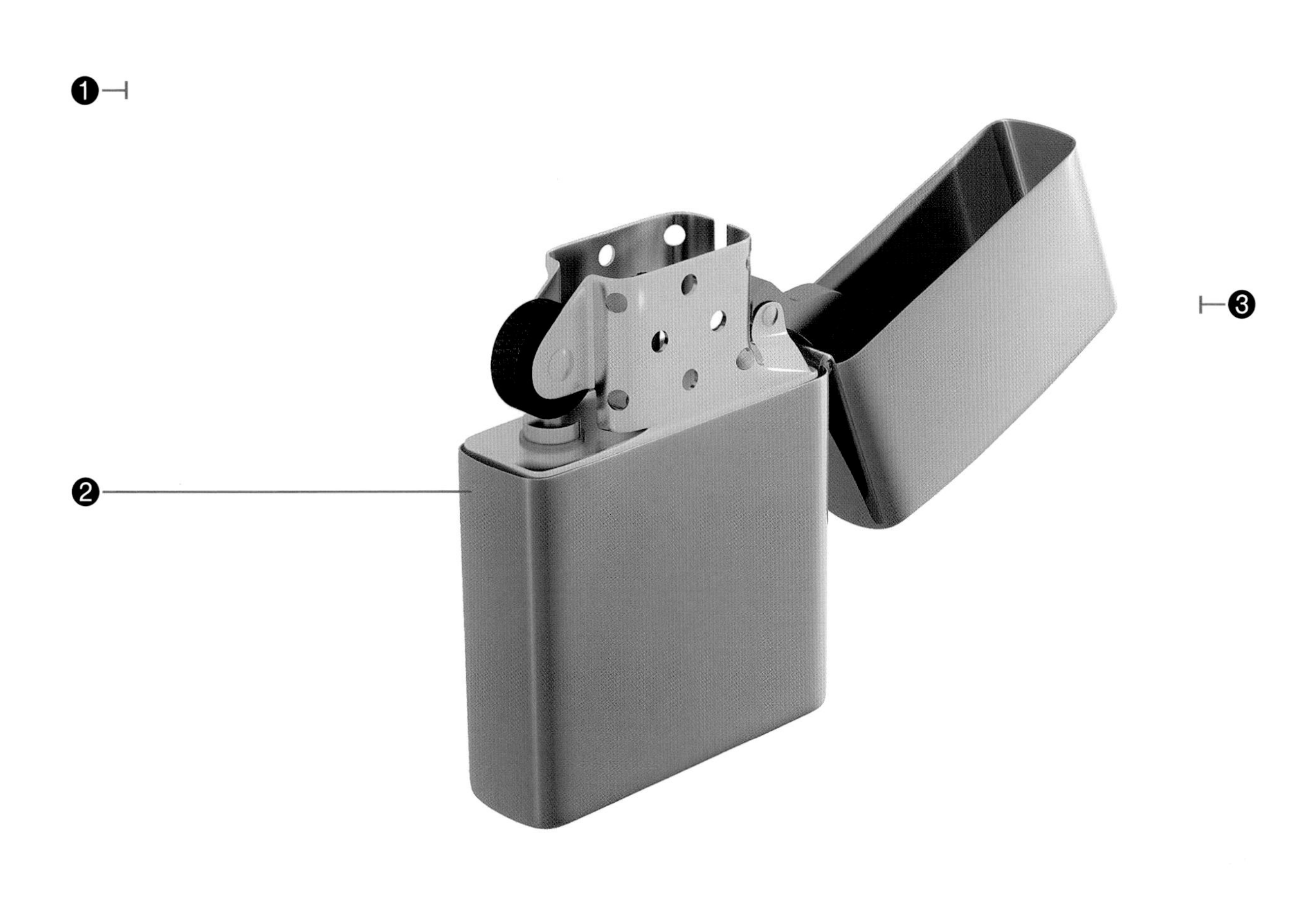

ZX81

Rick Dickinson and Sinclair Research, 1981

❶ Originally available as a mail-order kit only, the Sinclair ZX81 and its predecessor, the ZX80, were Clive Sinclair's attempts to bring computing to the masses. But the ZX81 was more than a kit computer. Its small size, sleek shape, and modern aesthetic combined with its low cost—it was the original "$100 computer"—to set design standards and price expectations for personal computing devices that influence to this day. Rick Dickinson comments: "Apart from Apple, the typical personal computer lacked true industrial design direction and I felt certain their correct identity was as yet undiscovered, but from where should a home computer take visual reference—a TV set, hi-fi, furniture, typewriters? Furthermore, the ZX81 was so much smaller, adding yet another layer of identity confusion. Sinclair Industrial Design aesthetics have historically been established around simple and pure forms, strict attention to detailing and perfect surface finishes, applied to the thinnest possible space envelopes, and above all a purist approach to general problem solving. For the ZX81 the starting point was the keyboard layout, so I established the smallest size the keys and key spacings could be based on the high quantity of data required per key and basic ergonomics in a predominantly graphic design process. This would set the width of the product. The PCB would site behind this to create the lowest possible profile. A slight angle to the keyboard and top surface added a more dynamic stance, and the rest was production process detailing."

❷ The ZX81 was the first designed computer—that is, the first computer to clearly follow a set of principles derived from an overarching design philosophy, which was unapologetically Bauhaus. The result: coherence and timelessness.

❸ The case design is modern and minimalist, with ports discreetly located on the side and back. Early prototypes explored pressed steel and aluminum, but ultimately moved to injection molded black plastic, due to problems with suppliers. The general aesthetic is viable even by contemporary standards, though the engineering was never up to the high expectations set by the design. The ZX81 sold 1.5 million units, until it was replaced by its successor in 1983, the ZX Spectrum, a more capable nonkit computer. As the Sinclair product line made the shift from low-cost DIY kit computer to fully functional off-the-shelf personal computer, it sacrificed its differentiation and product forgiveness. Facing rising costs and significant competition in the personal computer space, the Sinclair line of computers was discontinued in 1986.

❹ The membrane keyboard dramatically reduced the number of components in the computer, enabling its low price point. The keyboard is simple, elegant, spill resistant, but small and marginally functional, at least in terms of traditional typing. In terms of entry-level, hobbyist programming—the computer's intended function—the keyboard is perfectly adequate. Advanced users often bypassed the membrane keyboard by hacking traditional surplus keyboards to work with the ZX81. Interestingly, and perhaps a tribute to its low cost and DIY persona, there was a high degree of product forgiveness with these kinds of "upgrades"—the effect was generally positive, educational, creating little or no ill will toward the product itself. The keys, glyphs, and letters are high-contrast black and white. Functions and commands are white on black and red on white. Spacing, alignment, and color-coding are used to concentrate and organize an amazing amount of information on the keys with minimal visual noise. A case study in good graphic design.

1
2
ZX81
3
4

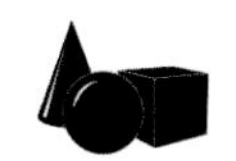

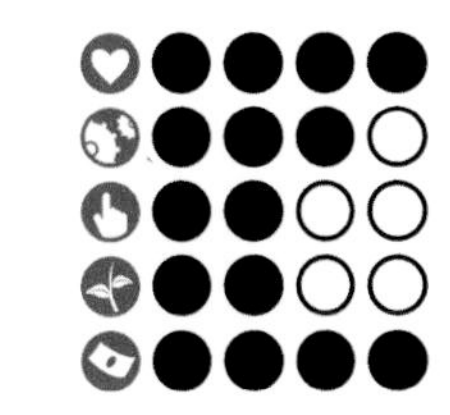

CONTRIBUTORS

ROBERT BLAICH INDUSTRIAL DESIGN

A U.S. industrial designer with a BFA in industrial design from Syracuse University, Robert Blaich worked in design and marketing before starting with Herman Miller Furniture Company in 1953, and served as its vice president of corporate design and communications from 1968 to 1979. In 1980, he became head of design at Royal Philips Electronics in the Netherlands, where he integrated engineering, marketing, and design and built a new corporate image of global design.

An active member of IDSA, he became a fellow in 1981. He was president of the International Council of Societies of Industrial Design (ICSID) from 1985 to 1987. In 1990, he received an honorary doctorate in design from Syracuse University and in 1991, was knighted by Queen Beatrix of the Netherlands.

Blaich remained at Philips until his retirement in 1992, when he founded Blaich Associates in Aspen, Colorado. He is author of *Product Design and Corporate Strategy* (1993), and *New & Notable Product Design* (1995).

JILL BUTLER GRAPHIC DESIGN

Jill Butler has been drawing, designing, and creating things for as long as she can remember. Some of her earliest memories include her mother's dispensing of design advice such as, "Everything looks better outlined in black" and "Odd numbers of items are more interesting than even numbers."

To test these theories, Butler worked as a designer and consultant for various organizations before founding Stuff Creators Design Studio in 2002. Since its inception, Stuff Creators has worked on a wide variety of projects including identity design, museum exhibit design, 3D modeling, website design, online training, and interactive design. Artistic creations include cows for the Houston Cow Parade and custom tables for the Chairs for Children fund-raiser.

Butler is coauthor of the book *Universal Principles of Design*. She has also taught design, desktop publishing, and typography classes at the University of Houston and Kingwood Community College. An award-winning graphic designer, Butler has interests ranging from information design and user experience design to illustration and paper engineering. She lives and works in Houston, Texas.

ALAN COOPER TECHNOLOGY DESIGN

Alan Cooper is a pioneer of the modern computing era. He is credited with creating what many regard as the first serious business software for microcomputers, and his groundbreaking work in software invention and design has influenced a generation of programmers, designers, and businesspeople and helped a generation of users.

For the last fifteen years, his interaction design consultancy, Cooper, has helped companies invent powerful, usable, desirable digital products via his unique methodology, Goal-Directed® Design. A cornerstone of this method, Personas, has been broadly adopted across the industry. Alan is the author of two industry best-selling books, *About Face* and *The Inmates Are Running the Asylum*, and is widely known as the "Father of Visual Basic."

BROCK DANNER ARCHITECTURE

Brock Danner is a designer and consultant based in New York. As an award-winning architect, and environment and broadcast designer Brock is passionate about user-based design innovation. He has applied his talents to the design of buildings, exhibitions, stores, furniture, products, brand identities, graphics, and communication programs. His credits include CNN's *Anderson Cooper 360°* and the worldwide CNN International studios, as well as consumer experience strategies and retail design for Wal-Mart, Samsung, and Wired Magazine. Brock has a master's degree in advanced architectural design from Columbia University, and served as adjunct professor of architectural design at the New Jersey Institute of Technology.

KIMBERLY ELAM GRAPHIC DESIGN

Kimberly Elam is a writer, educator, and graphic designer. She is currently the chair of the Graphic & Interactive Communication Department at the Ringling College of Art + Design, Sarasota, Florida. She has written and lectured extensively about graphic design and design education. Her first book, *Expressive Typography: Word as Image* (Van Nostrand Reinhold), identifies and analyzes methods by which words can transcend didactic meaning and become images. *Geometry of Design; Studies in Proportion and Composition* (Princeton Architectural Press) visually illustrates the connection between classic proportioning systems and modern graphic design, industrial design, illustration, and architecture. *Grid Systems: Principles of Organizing Type* (Princeton Architectural Press) puts forth a clear methodology for understanding and learning the grid system of composition. *Graphic Translation* focuses on the creation of an image with the visual means of abstraction, reduction, and interpretation with point, line, plane, shade, and shadow. Her most recent book, *Typographic Systems: Rules for Organizing Type* (Princeton Architectural Press), presents an innovative series of nontraditional, rule-based, visual language systems for typographic composition. Her current work focuses on the development of a series of innovative ebooks and print-on-demand books for design education on the website StudioResourceInc.com.

DONALD EMMITE DESIGN HISTORY

Don Emmite is a design specialist and consultant with thirty years' experience spanning several areas of the design field: interior design, industrial design, and graphic design. He earned a BFA from Sam Houston State University in advertising/graphic design. He is a registered interior designer certified by the National Council for Interior Design Qualification (NCIDQ) practicing in Houston, Texas, for twenty years.

Emmite's focus has shifted from interior design to that of industrial design historian through exhaustive journal and patent research allowing him to document important product designs that shaped American lifestyle in the middle of the twentieth century. For twenty-five years, he has been a collector of mid-twentieth-century American product design that includes classic modern furniture, Russel Wright designs, and appliances from the 1930s through the 1950s.

He has served as curator and designer of several exhibitions, most notable being Designing Domesticity: Industrial Design for Modern Living, 1930–1960, installed in the Gallery and Archives at the University of Houston Gerald D. Hines College Of Architecture in fall 2006. His home and collection have been featured in *The Houston Chronicle, Houston House and Home* magazine, *Home Fort Lauderdale Magazine*, and *Atomic Ranch* magazine.

LARIMIE GARCIA GRAPHIC ARTS

Most know him as Larimie. He was raised in the California sunshine by other mammals in concrete dwellings nestled in a land of a thousand oaks. Nurtured by elders of his tribe, he grew up to be an eclectic creature. Guided by the deer and powered by the wolf, his keen sensibilities reflect an intrigue and interest of the harmonies and anomalies that exist in nature. His visions are inspired by ancient and contemporary cultures as well as structures, patterns, and systems occuring in nature. Fragments of his experiences exist widely through media in the modern world.

Garcia works in a range of media without limitations. From paper to pixels, he creates graphics and imagery for television, film, print, the Internet, and fashion. He has packaged albums with compelling imagery for musicians known worldwide. His insights have been cultivated by companies seeking more fulfillment and excitement in their brand. His art can be seen outside of his cavern in galleries and on the grittiest urban streets. He continues to generate creative content and sublime art direction in new media and other visual experiences that surprise the imagination.

SCOTT HENDERSON PRODUCT DESIGN

Scott Henderson is an American designer who heads the New York–based design studio Scott Henderson Inc., and is also principal and cofounder of MINT, where he designs, manufactures, and distributes home accessory objects to over 450 retailers and museums throughout the world, including the Museum of Modern Art and Design Within Reach.

Henderson is known for his ability to transform the mundane. He believes that if something makes you smile, it becomes easier to use. With over fifty patents in the U.S. and Europe—for projects as diverse as housewares and home accessories to consumer medical products, electronics, even aircraft—his work has been widely recognized in exhibitions, awards programs, and the press. For example, Scott's work for the juvenile products brand Skip Hop was recently featured in INC Magazine's special issue, *The Inc. 500*, listing the top 500 privately held businesses and specifically citing Scott Henderson's Skip Hop products as helping the company to achieve 786 percent growth. Included in Scott's career are multiple IDEA Awards—considered the most prestigious design award in the United States and given by the Industrial Designer's Society of America and *Business Week* magazine.

Henderson's work is included in the permanent collections of The Brooklyn Museum, the Chicago Athenaeum and Cooper Hewitt National Design Museum/ Smithsonian Institution.

KRITINA HOLDEN HUMAN FACTORS

Kritina Holden is the technical lead and acting manager of the JSC Usability Testing and Analysis Facility (UTAF). She is responsible for proposal writing, project planning, personnel tasking and mentoring, and ensuring technical quality of UTAF products. She also serves a technical role, providing human factors support to the Orion Crew Exploration Vehicle Cockpit team in activities such as requirements development, task analysis, and crew evaluations. Holden received her PhD in engineering psychology from Rice University, with an emphasis in human-computer interaction. She has been with Lockheed Martin at JSC for approximately sixteen years, in addition to two years with a Web-based training company, and three years working for a large, international software company. Holden is also an adjunct professor at the University of Houston-Clear Lake.

ROBERT KINGSLYN GRAPHIC DESIGN

Robert Kingslyn, a self-confessed jack-of-all-trades, has parlayed a communications degree into a meandering career path from corporate training, to software development, to e-books, including several liaisons with that fickle mistress, entrepreneurship. Consistent throughout has been a commitment to clear, simple design—with a particular focus on graphic design, branding, and the development of Web-based user experiences. Kingslyn's design career benefits from positions with government, corporate, and startup entities, each contributing a unique set of limitations, challenges, and opportunities. His broad range of skills and experiences has led Kingslyn to be described as a "'big D' designer" able to synthesize a wide range of complexities resulting in "simple and elegant solutions." Kingslyn currently provides contract design services to select clients while remaining hell-bent on a low-paying professional career as author/curmudgeon/world traveler. He considers himself fortunate to be married, have three dogs, and live amid the earthquakes, fires, and beautiful weather of Southern California.

JON KOLKO INTERACTION DESIGN

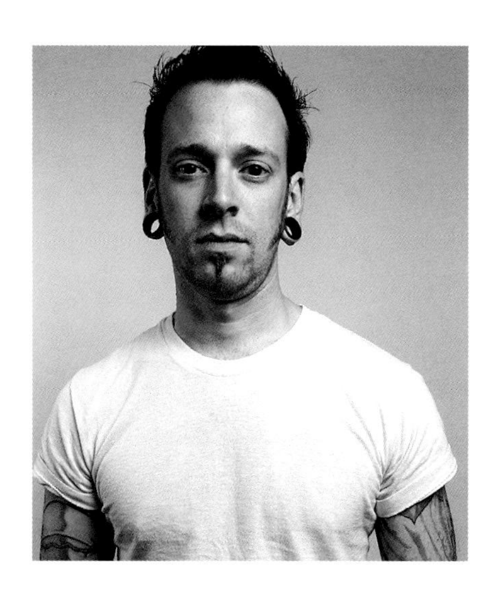

Jon Kolko is a senior design analyst at frog design. He has worked extensively in the professional world of interaction design, working around complicated technological constraints to best solve the problems of Fortune 500 clients. His work has extended into the worlds of supply chain management, demand planning, and customer-relationship management, and he has worked with clients such as Bristol-Myers Squibb, Ford, IBM, Palm, and other leaders of the Global 2000. The underlying theme of these problems and projects was the creation of a solution that was useful, usable, and desirable. His present research investigates the process of design, with a focus on the synthesis phase of the design process.

Prior to frog, Kolko was a professor of interaction and industrial design at the Savannah College of Art and Design, where he was instrumental in shaping the Interaction and Industrial Design programs. He is the coeditor in chief of *interactions* magazine, author of the text *Thoughts on Interaction Design*, and has been frequently published in the proceedings of ACM SIGCHI, IDSA, and the European Academy of Design conferences.

LYLE SANDLER EXPERIENCE DESIGN

Lyle Sandler is a design and innovation consultant. Previously, he served as director of the Advanced Concepts Group for Merrill Lynch. Lyle consults with numerous organizations in the area of experience architecture and the effects of multisensory stimuli in trade environments. He has developed solutions for the InterPublic Group, Publicis Groupe, Reuters, Goldman Sachs, AT&T, Dupont, Samsung, and Macy's, as well as the governments of Mexico, Indonesia, and Jordan.

ROB TANNEN HUMAN FACTORS

Rob Tannen is a human-centered design researcher with expertise in both products and user interfaces. He has over fifteen years' experience in designing and evaluating usable systems for organizations including Microsoft, the New York Stock Exchange, Siemens, and the United States Air Force.

Tannen is director of research at Bresslergroup, a leading design and development firm, where he works on a range of medical, industrial, and consumer products.

Tannen is also a professional writer and presenter on topics ranging from simplicity in design to effective research methods. He created DesigningforHumans.com, a reference blog for designers. He has also represented the human factors perspective as a judge for several major design awards including the Human Factors Society Product Design Group, IDSA International Design Excellence Awards, *I.D.* magazine's Annual Review, and Consumer Electronics Association.

Tannen is currently vice chair of the Human Factors section of the Industrial Designers Society of America (IDSA). He earned a PhD in human factors from the University of Cincinnati and is a certified professional ergonomist.

DORI TUNSTALL DESIGN ANTHROPOLOGY

Dori Tunstall is an associate professor of design anthropology and associate dean of learning and teaching at Swinburne University in Melbourne, Australia. She is a leader in the field of design anthropology: an emerging field that seeks to understand how the processes and artifacts of design help define what it means to be human.

She is passionate about civically engaged design that creates politically informed and enfranchised people. Her projects on Design + Governmentality have targeted voting and emergency evacuation with the organization, Design for Democracy, and independent projects on public health care and IRS design management. She is organizer of the U.S. National Design Policy Summit and Initiative.

Previously, Tunstall worked at Sapient Corporation and then Arc Worldwide as a senior experience planner. She spent five years bringing ethnographic insight to strategic design planning and implementation. Her roster of clients included: U.S. Army and Army Reserve, Northern Trust Bank, Sears, Whirlpool, Fujisawa, Unilever, Allstate Financial, and Nokia.

Dori holds a PhD and MA in anthropology from Stanford University and a BA in anthropology from Bryn Mawr College.

STEVEN UMBACH PRODUCT DESIGN

Steven Umbach founded the Texas-based Umbach Consulting Group in 2001, focused on product design innovation and new product development (NPD). Umbach earned a bachelor of fine arts in industrial design from the Rhode Island School of Design and a master's of science in management with emphasis on design management from the Georgia Institute of Technology.

His product design career is marked by innovation and significant product design firsts, including the first rumen injection device for deworming cattle, the first sonar-controlled stair-climbing wheelchair, the enclosure design for the first electronic keyboard that accurately recreates the sound of a grand piano, the first vacuum cleaner that converts from an upright into smaller handheld, the first mini-circuit vacuum-assisted prepackaged heart bypass system, and complete content and process development for the first green technology product design competition offered by a major computer manufacturer.

His early-career design work earned peer recognition through several Industrial Design Excellence Awards (IDEA) and a "10 Best Designs of the Decade" award from the Industrial Designer's Society of America (IDSA) for his comprehensive design of the injection device for cattle. Since 2003, Umbach has also served as adjunct faculty for the University of Houston, Gerald D. Hines College of Architecture, where he teaches human factors and ergonomics and professional business practice in the industrial design department.

PAULA WELLINGS INTERACTION DESIGN

Throughout her career, Paula Wellings has pursued her interest in the situative and psychosocial nature of how people interact with artifacts and environments. She advocates that great experiences require thinking strategically about the ecological systems formed by people, organizations, markets, and stakeholders and delivering products and services that evolve these systems in positive ways.

Past employers and clients include Human Code, Sapient, projekt202, Stanford University, WOWIO, Hasbro Interactive, IC2 Institute, Texas Workforce Commission, Dell, Citibank, Teachscape, Emily Carr Institute of Art and Design, Spark Media, Thomson West, and Microsoft.

Paula received her master's of art in earning, design and technology from Stanford University and her bachelors of design in electronic communication design from the Emily Carr Institute of Art and Design.

KHRISTINE ANDERSON

GRAPHIC DESIGN

Khristine Anderson is a designer for higher education. Days are spent designing websites and print pieces; creating marketing materials; enhancing websites with multimedia, animation, and video; and dabbling in social media. She enjoys running after a three-year-old, playing complicated board games, and capturing the moment with a digital camera.

MASSAD AYOOB

FIREARMS AND SELF-DEFENSE

Massad Ayoob is an internationally known firearms and self-defense instructor. He is the director of the Lethal Force Institute in Concord, New Hampshire, has taught police techniques and civilian self-defense to both law enforcement officers and private citizens in numerous venues since 1974, and has appeared as an expert witness in several trials. He has served as a part-time police officer in New Hampshire since 1972.

Ayoob has authored several books and over one thousand articles on firearms, combat techniques, self-defense, and legal issues, and has served in an editorial capacity for *Guns Magazine, American Handgunner, Gun Week, and Combat Handguns*. Since 1995, he has written self-defense and firearms-related articles for *Backwoods Home Magazine*. He also has a featured segment on the television show *Personal Defense TV*, which airs on The Outdoor Channel in the United States.

MARVIN AYRES

MUSIC

A founding member of the Government, Marvin Ayres has also made solo contributions to bands and artists throughout the late 1980s and 1990s, notably, Culture Club, Simply Red, Frankie Goes to Hollywood, and Prefab Sprout. In 1999 he was signed to the avant-garde label Mille Plateaux as composer/producer, and released two ambient albums, *Cellosphere* and *Neptune*, to critical acclaim.

His reputation as producer and performer came to the attention of Jaki Liebezeit (ex Can), who invited him to remix tracks and tour with Liebezeit's band, Club off Chaos, in early 2001. Ayres is currently commissioned by Einsturzende Neubauten's electronica music publishers Freibank, and his compositions have been featured on five of Freibank's "For Films" albums.

JONTI BOLLES

ARCHITECTURE

Jonti Bolles is currently responsible for overseeing search engine marketing strategy (SEM) for Schipul and the company's clients. She regularly works with UI experts, programming teams, and clients to achieve a blended solution for web design, applications, and search engine strategies.

Prior to joining Schipul, Bolles taught architecture at Prairie View A&M University and University of Houston and became enamored with the convergence of design and technology. She also studied instructional design and worked at Houston companies Eduneering and MindOH! as a project manager, developing online educational tools for corporate compliance, and a product development manager, reaching at risk youth in a technology space.

Bolles is a graduate of Texas Wesleyan University with a BBA in economics/finance and received her master's of architecture from University of Texas at Arlington. Bolles is a native Texan who loves traveling outside the United States and proudly defends her choice of a Ducati for daily transportation. Bolles can easily be distracted by the newest shiny tech and design objects and social media theory.

CARLY FRANKLIN

GRAPHIC DESIGN

Carly Franklin is principal of CFX Creative, a Vancouver-based design studio specializing in corporate identity and branding work. Since founding the studio in 1999, Franklin has worked on a wide range of identity, print, packaging, and web projects, with clients throughout the United States and Canada. She is an active professional member of the Society of Graphic Designers of Canada.

A self-taught designer, Franklin credits the years she spent working in art galleries and custom frame shops as the foundation of her design education. Early in her career, she fell in love with identity design after a chance encounter with a stack of design books at the local public library. Originally a native of Baton Rouge, Louisiana, she currently lives and works in Vancouver, BC.

JENNIFER GARRETT

INDUSTRIAL DESIGN

Jennifer Garrett is a young industrial designer currently working in the field of audio consumer electronics. Several of her student and professional works have been featured in various print and online publications or shown at trade events such as the 2007 International Contemporary Furniture Fair, the 2008 Industrial Designers Society of America Northeast District Conference, and the 2008 and 2009 Consumer Electronics Shows.

Garrett is an active contributor to the growing design community and is a professional member of the IDSA in the Northeast. At the time of this publication, Garrett is employed by New Jersey–based SDI Technologies to design products and packaging under the international brand names of Timex and iHome, the number one market leader in iPod/iPhone electronics in North America, and leisurely pursues her creative hobbies from her home in southern New Jersey.

DAVID KNAUB

MECHANICAL ENGINEERING

Dave Knaub has been a senior mechanical engineer at Ziba Design of Portland, Oregon, for over twenty years. He holds numerous patents and design awards for his work in the design of products, mechanisms, structures, and electromechanical devices. He previously worked as a mechanical engineer at Texas Instruments Inc. and has also been an instructor at the Art Institute of Portland.

Knaub received bachelor's and master's degrees in mechanical engineering from Rensselaer Polytechnic Institute, and a master's in engineering product design from Stanford University. A registered professional engineer, he is a member of the American Society of Mechanical Engineers and the Society of Plastics Engineers, and is an accomplished symphonic contrabassist.

VLAD KUNKO

INDUSTRIAL DESIGN

A Boeing 777, Lunstead/Haworth executive furniture, a NASA/DoE Wind Turbine, Magic apparel show exhibits, and a USIA Industrial Design Specialist-in-Residence with Design USA all have Vlad Kunko in common.

Today, Vlad is a meta-designer specializing in design research, social and cultural cybernetics, psycho-ethnography, semiology, and design history. His design career began as a model maker with NASA and developed to include furniture and exhibit design. His graduate work in whole systems design furthered his interest in cognition, design process, and ecological imperatives.

With seven years in academia, he currently serves as director and faculty at the Antropos Design Institute, Edmonds, Washington. He also manages the AnDi Design Collection as curator and historian. www.antroposdesign.org

KEITH LANG

INTERACTION DESIGN

Keith Lang is an interaction designer who works to make everyday computing easier and more fun. Lang cofounded Plasq.com, an independent software company that created the highly popular and award-winning titles *Comic Life* and *Skitch*. Software design, the history of computing, and cognitive psychology are all topics covered in his blog at UlandUs.com—he'd love for you to visit and share your thoughts.

ROBERT MOORE

FOOT AND ANKLE SURGERY

Robert Moore has more than twenty years of experience in the health, fitness, and medical fields. He was trained in Houston in foot and ankle surgery in the early 1990s, and established his private practice in 1993, specializing in corrective surgery, trauma, and sports medicine. He now owns and operates three clinics in the Houston area.

Moore is author of the book, *Body of Knowledge,* founder and chief executive officer of Body of Knowledge, Inc., and a member of the Mayor's Wellness Council in Houston, Texas.

JAMES MUELLER

INDUSTRIAL DESIGN

Jim Mueller is an industrial designer who specializes in assistive technology, disability management, and universal design. Since 1974, Mueller has served as a consultant to employers, product manufacturers, and rehabilitation research centers. He has also designed and fabricated hundreds of workplace and home modifications for individuals with disabilities.

Mueller helped establish IDSA's Special Interest Section on Universal Design in 1994. Taking over as section chair in 1997, he helped grow this group from 24 members to more than 500 by 2009. From 1994–2004, Mueller served as a consultant to North Carolina State University's Center for Universal Design, where he helped establish the 7 Principles of Universal Design.

DAVID PALUMBO

INSTRUCTIONAL TECHNOLOGY

David Palumbo is a frequent speaker, regarding information management, adoption, and enterprise learning initiatives. He has led discussions on the business of learning and the role of emerging technologies at various universities around the globe. He has forty-eight writings to his name in publications such as *Computers in Human Behavior, Computers in the Schools, The Journal of Educational Issues of Language Minorities Students*, and *Journal of Computing in Childhood Education*. Palumbo has also received forty-one industry awards for the design and development of interactive software projects.

Palumbo holds a doctorate in educational psychology and instructional technology and a master's degree in educational psychology from West Virginia University. He also holds a bachelor's degree in psychology from the University of Michigan.

UMESH PERSAD

ENGINEERING DESIGN

Umesh Persad is an assistant professor in the Product Design and Interaction Lab, Centre for Production Systems at the University of Trinidad and Tobago (UTT). He researches the creation of products and interfaces that can improve health care and quality of life for older and disabled users by adopting inclusive/universal design. He is also a member of the IEEE Engineering in Medicine and Biology Society (EMBS). In addition to his academic work, Persad is also a trained Web/graphic designer and a performing musician. He is a member of the Theological Board and Director of Publications for Swaha International, a Hindu-based nonprofit organization working toward the spiritual, social, and cultural development of Trinidad and Tobago.

STEVE PORTIGAL
DESIGN STRATEGY

Steve Portigal is the founder of Portigal Consulting, a San Francisco Bay Area firm that brings together user research, design and business strategy. Portigal Consulting helps clients to discover and act on new insights about themselves and their customers. In addition to regularly speaking at design and marketing events, Portigal has taught design research at the California College of Art and the Involution Master Academy. He writes regularly for Core77.com and the Portigal Consulting blog, www.portigal.com/blog. Portigal is an avid photographer who has a Museum of Foreign Grocery Products in his home.

DAN SAFFER
INTERACTION DESIGN

Dan Saffer is a founder and principal designer at Kicker Studio. He has designed devices, software, websites, and services since 1995, and these products are currently used by millions every day. An acclaimed speaker and author, his two books are *Designing for Interaction* (New Riders) and *Designing Gestural Interfaces* (O'Reilly). His design innovations have received several patents.

Saffer is an internationally recognized thought leader on design who has spoken at conferences and taught workshops on interaction design all over the world. He also has the distinction of coining the phrase "topless meeting." which was a finalist for Oxford's Word of the Year and *Time* magazine's number ten Buzzword in 2008. He has a masters of design in interaction design from Carnegie Mellon University.

BUDD STEINHILBER
INDUSTRIAL DESIGN

Budd Steinhilber began his design apprenticeship with Raymond Loewy Associates in 1942. He was the first employee of Dohner & Lippincott, which was later to become Lippincott & Margulies. He was part of the L&M design teams that helped develop the 1948 Tucker rear-engine automobile, as well as the planning of the crew and officers' quarters for America's first nuclear submarine, the USS *Nautilus*.

Supposedly "retired," he has spent much of the past decade involved in the design development of electric vehicles and battery systems. He lives on the Big Island of Hawaii.

DANNY STILLION
INTERACTION DESIGN

Danny Stillion is a design director and associate partner at IDEO. With a background in fine art, graphic design, and product design, Stillion was drawn to the field of interaction design through his early interests in communication design and time-based media and how we physically interact with products and services. Prior to joining IDEO, Stillion taught interactive media courses in higher education and consulted as an interaction designer. Today he continues to mentor and learn from other design colleagues in the context of active projects that span telecommunications, medical, service design, consumer products, and vehicle telematics domains. His work and perspective have been featured in the Museum of Modern Art (New York), *I.D. Magazine*, *BusinessWeek*, *ACM Interactions*, and the IDSA Industrial Design Excellence Awards.

ROBIN TAYLOR

COMPETITIVE SHOOTING

Robin Taylor has been immersed in semi-pro shooting for more than twenty years. His expertise rises from the rough-and-tumble contests favored by civilian and military marksmen, giving him a penetrating real-world expertise on firearms, particularly pistols. Taylor is a Glock Shooting Sports Foundation Master, Glock Armorer, staffer at the United States Practical Shooting Association (USPSA), assistant editor for USPSA's *Front Sigh*t magazine, and author of *The Glock in Competition*. He is founder of Taylor Freelance, LLC, makers of competition firearm accessories.

SHANNON THOMAS

USER

Shannon Thomas is an artist living in Portland, Oregon. When she was eight years old, Thomas was in a car accident, leaving her paralyzed from the chest down. She began using an iBot in 2007 and writes about her experiences at www.wheelchairrevolution.blogspot.com.

TREVOR VAN GORP

USER EXPERIENCE DESIGN

Trevor van Gorp is the founder and principal of Affective Design Inc., an independent user-experience consultancy. He has been working in design and visual communication since 1994 and holds a BFA in graphic design and a master's of environmental design in industrial design, specializing in human computer interaction.

He is the author and editor of affectivedesign.org, a blog dedicated to exploring and understanding the "heart" of design: the effects of design on the emotional affect created by people's interactions with products, brands, and services.

Van Gorp is a member of the Information Architecture Institute, the Association of Canadian Ergonomists, and the Design & Emotion Society. He has given presentations on the topic of design and emotion at conferences both in North America and internationally.

JASPER VAN KUIJK

INDUSTRIAL DESIGN

Jasper van Kuijk is a PhD candidate at the Faculty of Industrial Design Engineering (TU Delft), studying usability in the development of electronic consumer products. The aim of his study is to determine what factors in product development practice determine the usability of electronic consumer products, such as mobile phones, mp3 players, and washing machines.

Van Kuijk is also the author of "the product usability weblog" (www.uselog.com) where he writes about consumer product usability. On this weblog you'll find examples of products with good and poor usability, usability studies, summaries of interesting papers, and news and events.

At the faculty of Industrial Design Engineering, van Kuijk works in the ID-StudioLab, a hotbed of design researchers, research designers, design educators and a lot of other hybrids. Besides doing design research, van Kuijk is a (semi-professional) stand-up comedian.

❶ **ISAAC AND HENNA SHAH** RESEARCH, USABILITY TESTING

❷ **SCOTT O'CONNOR** PRODUCT MODELING

❸ **TY, JENNIE, JACKLYN, KIP, AND OLIVIA HELM** USABILITY TESTING

❹ **ROSS MASON** RESEARCH

❺ **LUKE GARRISON** USABILITY TESTING

❻ **FRITZIMUS "FRITZ" MAXIMUS** USABILITY TESTING

❼ **RANDY COLLIER** IMAGE PROCESSING

❽ **DAVE SEVIGNY** SUBJECT-MATTER EXPERT

❾ **AARON AND JAMIE WARREN** USABILITY TESTING

❿ **JASON HOLDEN** RESEARCH

REFERENCES

9093 Kettle, quotation from "Michael Graves: Beyond Kettles" by Pallavi Gogoi, *BusinessWeek*, August 18, 2005.

Absolut Vodka Bottle, quotation from *Absolut: Biography of a Bottle* by Carl Hamilton, Texere, 1994.

Aeron Chair, quotation from "When Choosing Chairs, Take Your Office Attitude Home With You" by Katherine Salant, The *Washington Post*, May 21, 2005.

AIBO ERS-110, quotation from *AIBO Town*, July 2000.

AJ Cutlery, quotation from *Arne Jacobsen* by Christopher Mount, Marisa Bartolucci, and Raul Cabra, Chronicle Books, 2004.

Angled Measuring Cup, quotation from Alex Lee, *Gel Conference*, 2008.

Ball Clock, quotation from *George Nelson: The Design of Modern Design* by Stanley Abercrombie, MIT Press, 2000.

Ball Vacuum Cleaner, quotation from "A Conversation with ... James Dyson" by Allan Chochinov, *Core77*, October 2002.

BeoCom 2 Phone, quotation from "Telecom: Beocom 2," www.bang-olufsen.com, Fall 2002.

Bratz Doll, quotation from "How Bratz Beat Barbie" by David Rowan, *The Times Magazine*, December 4, 2004.

Bubble Lamp, quotation from *George Nelson: The Design of Modern Design* by Stanley Abercrombie, MIT Press, 2000.

Cabbage Patch Kids, quotation from "Doll Creator: Coming Up with Fresh Ideas Will Be the Challenge" by Jamie Gumbrecht, *The Atlanta Journal Constitution*, September 20, 2008.

Cameleon Stroller, quotation from "Five-Minute Time Out: Max Barenbrug" by Gwynne Watkins, www.babble.com, March 30, 2007.

City Hall Clock, quotation from *Jacobsen* by Felix Solaguren-Beascoa, Santa & Cole, 2002.

ClearRx Pill Bottle, quotation from "Clear RX: Design on Drugs" by Evelyn Hafferty, Brandchannel.com, September 5, 2005.

Coca-Cola "Contour" Bottle, quotation from *Marketing Communications* by P. R. Smith and Jonathan Taylor, Kogan Page Publishers, 2004.

Coldspot Refrigerator, quotation from *Never Leave Well Enough Alone* by Raymond Loewy, John Hopkins University Press, 1951.

Crescent Adjustable Wrench, quotation from *Cooper Industries: 1833–1983* by David Neal Keller, Ohio University Press, 1983.

Cristal Ballpoint Pen, quotation from "Inspired by Wheelbarrow, BIC Sells 100 Billionth Pen" by Tim Hepher, *Reuters*, September 8, 2005.

Diamond Chair, quotation from *The World of Bertoia* by Nancy Schiffer and Val Bertoia, Schiffer Publishing, 2003.

Dish Soap Bottle, quotation from *Karim Rashid: Compact Design Portfolio* by Marisa Bartolucci and Raul Cabra, Chronicle Books, 2004.

Dixon Ticonderoga Pencil, quotation from "All About Our Company," www.dixonusa.com, March 2009. See also *The Pencil: A History of Design and Circumstance* by Henry Petroski, Knopf, 2006.

Egg Bird Feeder, quotation from correspondence between Jim Schatz and author.

Forehead Thermometer, quotation from correspondence between Scott Henderson and author.

FuBar Functional Utility Bar, quotation from "Please Hammer, Hurt 'Em" by Joe Brown, *I.D.* magazine, September 01, 2006.

Garbino Trash Can, quotation from "Design Icon" by Bettijane Levine, *Los Angeles Times*, March 13, 2008.

Glock 17 Pistol, quotation from "Top Gun" by Dyan Machan, *Forbes*, Mar 31, 2003.

Good Grips Peeler, quotation from "At the Heart of Interaction Design" by Lauralee Alben, *Design Management Journal*, Summer 1997.

HomeHero Kitchen Fire Extinguisher, quotation from "Getting Warmer" by Eva Hagberg, *Metropolis*, October 2007.

Hug Salt and Pepper Shakers, quotation from correspondence between Alberto Mantilla and author.

HurriQuake Nail, quotation from "Dr. Nail" By Susan Polowczuk, *Clemson World*, Summer 2007.

iBOT Mobility System, quotation from "The Dean of Engineering" by William Lidwell, *Make*, 04, 2005.

il Conico Kettle, quotation from *A Scientific Autobiography* by Aldo Rossi, MIT Press, 1981.

IN-50 Coffee Table, quotation from *A Sculptor's World* by Isamu Noguchi, Harper & Row, 1968.

iPhone, quotation from "The Boy from Chingford Who Puts the Bite into Apple's Iconic Design" by Claire Beale, *The Independent*, May 19, 2008.

iPod 5G, quotation from "The Guts of a New Machine" by Rob Walker, The *New York Times*, November 30, 2003.

Jaipur Foot, quotation from "P. K. Sethi, 80, Invented Prosthetic Foot" by Stephen Miller, *The Sun*, January 8, 2008.

Jawbone Wireless Headset, quotation from "Yves Behar: Creating Objects That Tell Stories," *TED*, February 2008.

Jelly Fish Watch GK 100, quotation from "Message and Muscle: An Interview with Swatch Titan Nicolas Hayek" by William Taylor, *Harvard Business Review*, March–April 1993.

Joystick CX40, quotation from "Interview: Bushnell's NeoEdge Becomes Big Player with Yahoo! Games" by John Gaudiosi, *GameDaily*, July 10, 2008.

Juicy Salif, quotation from "Starck Speaks: Politics Pleasure Play" by Philippe Starck, *Harvard Design Magazine*, Summer 1998.

LC4 Chaise Longue, quotation from *The Lc4 Chaise Longue* by Le Corbusier, Pierre Jeanneret, and Charlotte Perriand, Verlag Form, 1998.

LED Watch PH-1055, quotation from "Designer Philippe Starck Embraces His Feminine Side" by Linda Tischler, *Fast Company*, January 26, 2009.

Lil' Pup Dog Bowl, quotation from "Design of the Times: Everyday Objects Become Must-Haves, Thanks to Our New Love Affair with Style," *Austin American Statesman*, August 12, 2004.

Lounge Chair and Ottoman, quotation from "Norton Lecture 1" by Charles Eames, Harvard University, 1970.

Macintosh, quotation from "More Like a Porsche" by Andy Hertzfeld, www.folklore.org.

Maglite Flashlight, quotation from "Anthony Maglica: Mag Instrument" by Kemp Powers, *Fortune Small Business*, September 1, 2004.

Model 302-F1 Telephone, quotation from *Designing for People* by Henry Dreyfuss, Simon and Schuster, 1955.

MoneyMaker Pump, quotation from "Global X Interview: Martin Fisher—Kickstart," www.socialedge.org, 2008.

Mouse M0100, quotation from "The Making of the Mouse" by Alex Soojung-Kim Pang, *American Heritage* 17, no. 3, Winter 2002.
Museum Watch, quotation from *American Design Classics* by Nada Westerman and Joan Wessel, Design Publications, 1985.
Natural Nurser Baby Bottle, quotation from "A Fine Bottle: Whipsaw's New Design Is a More Natural Way for Babies to Nurse" by Lara Kristin Lentini, *Metropolis*, December 2007.
No. 5 Flacon, quotation from "No 5," www.chanel.com, January 2009.
0-Series Scissors, quotation from "About Fiskars: R&D," www.fiskars .com, January 2009.
PalmPilot, quotation from "Jeff Hawkins: The Man Who Almost Single-Handedly Revived the Handheld Computer Industry" by Shawn Barnett, *Pen Computing*, April 2000.
Pencil Sharpener, quotation from *Never Leave Well Enough Alone* by Raymond Loewy, Simon & Schuster, 1951.
Phonosuper SK 4, quotation from *Less But Better* by Dieter Rams, Jo Klatt Design + Design Verlag, 1995.
Pleo, quotation from "Maker: I, PLEO: How Caleb Chung Went from Street Mime to Toy-Robot Maven" by Robert Luhn, *Make*, 08, 2006.
Pocket Survival Tool, quotation from "Outdoing the Swiss Army Knife" by Fawn Fitter, *Fortune Small Business*, October 5, 2007.
POM Wonderful Bottle, quotation from "Spotlight: Glass Packaging: One-of-A-Kind Glass Bottle Crystallizes Branding Vision for POM Wonderful Juices" by Mary Ellen Reis and Jung Weil, *Package Design Magazine*, March 2004.
Post-It Note, quotation from "How We Got Stuck" by K. Prabhakar, *Tales of Organizations: 3M Case Study*, 2007.
Pot-in-Pot Cooler, quotation from "Global X Interview: Mohammed Abba—Nigeria," www.socialedge.org, 2007.
Power Mac G4 Cube, quotation from "The G4 Cube Introduction" by Steve Jobs, *Macworld*, New York, 2000. See also "When Apple Failed" by Andy Greenberg, *Forbes*, October 30, 2008.
Q-Drum, quotation from "The Q Drum—Water Transporter for Developing Countries" by Pieter Hendrikse, Changemakers.net, 2007.
RAZR V3 Phone, quotation from "Weaving Design into Motorola's Fabric: Interview with Jim Wicks" by Brandon Schauer, *Strategy '06 Conference*, May 17, 2006. See also "Moto's Mojo" by Chuck Salter, *Fast Company*, September 11, 2008.
Remote Control—Series 1, quotation from "Story of a Peanut: The TiVo Remote's Untold Past, Present, and Future" by Christopher Mascari, *GIZMODO*, June 20, 2008.
Roomba Robotic Vacuum Cleaner, quotation from "Is There a Robot in Your Future? Helen Greiner Thinks So," *Knowledge@Wharton*, April 5, 2006.
Round Thermostat, quotation from *Henry Dreyfuss: Industrial Designer: The Man in the Brown Suit* by Russell Flinchum, Smithsonian Institution, 1997.
Rubik's Cube, quotation from "The Perplexing Life of Erno Rubik" by John Tierney, *Discover*, March 1986.
Safety Razor, quotation from "The Razor King" by Howard Mansfield, *Invention and Technology Magazine*, Spring 1992.
Segway i2, quotation from "The Dean of Engineering" by William Lidwell, *Make*, 04, 2005.
Selectric Typewriter, quotation from *Eliot Noyes: A Pioneer of Design and Architecture in the Age of American Modernism* by Gordon Bruce, Phaidon, 2006.
Silent Cello SVC-200, quotation from *Design Secrets: Products: 50 Real-Life Projects Uncovered* by IDSA, Rockport Publishers, 2001.
Sixtant Electric Shaver, quotation from *Less But Better* by Dieter Rams, Jo Klatt Design + Design Verlag, 1995.
Soda King Siphon, quotation from *Horizons* by Norman Bel Geddes, Dover Publications, 1977.
Sonicare Toothbrush, quotation from conversation between David Giuliani and author.
Stand Mixer Model K, quotation from "Industrial Humaneer" by Don Romero, *Mechanix Illustrated*, December 1946.
StarTAC Mobile Phone, quotation from "StarTac" by Michael Köttl, *Mobile Times*, July 1996.
SX-70 Camera, quotation from *Insisting on the Impossible: The Life of Edwin Land* by Victor McElheny, Perseus Books, 1998.
THUMP 2 Sunglasses, quotation from conversation between Colin Smith and author.
Titanium Series Padlock, quotation from "Design Strategy and Strategic Design at Master Lock" by Gianfranco Zaccai, *Design Management Journal*, Winter 2002.
Tizio Table Lamp, quotation from *Tizio Light* by Richard Sapper, Birkhäuser, 2003.
TMX Tickle Me Elmo, quotation from correspondence between Ron Dubren and author.
TOUCH! Tableware, quotation from "Flock Fusion" by Wiebke Lang, *form: The European Design Magazine*, July–August 2005.
Transistor Radio TR-610, quotation from "The Creator of Our Private Universe" by Peter Popham, *The Independent*, March 22, 1996.
Variable Balans Stool, quotation from "People: Peter Opsvik," *ONOFFICE*, August 2008.
Vespa 150 GS, quotation from *Vespa: 1946-2006: 60 Years of the Vespa* by Giorgio Sarti, MotorBooks/MBI Publishing Company, 2006.
Walkman TPS-L2, quotation from *Made in Japan: Akio Morita and Sony* by Akio Morita, Edwin Reingold, and Mitsuko Shimomura, Collins, 1987.
Wiggle Chair, quotation from "Frank Gehry Interview," *Academy of Achievement*, June 3, 1995.
Wii, quotation from "The Engadget & Joystiq Interview: Nintendo's Shigeru Miyamoto (Again!)" by Ryan Block, www.engadget.com, May 11, 2006.
XO-1 Laptop, quotation from "Nicholas Negroponte: The Interview" by Wade Roush, *Xconomy | Boston*, January 28, 2008.
Zippo Cigarette Lighter, quotation from "Corporate Information: George G. Blaisdell," www.zippo.com.
ZX81, quotation from correspondence between Rick Dickinson and author.

INDEX

Note: Page numbers in italics indicate images.

ACKNOWLEDGMENTS

The authors would like to thank the many contributors who offered commentary, both by direct invitation and via the website. Even though only a small percentage of the overall commentary received could be included, all of it was appreciated and considered in our analysis. We would also like to thank the many designers and manufacturers who shipped us their products for testing and evaluation. Special thanks to Kristin Ellison, David Martinell, John Gettings, Tiffany Hill, Marcy Slane, Leslie Haimes, Allison Hodges, Winnie Prentiss, and Karina Descartin Manacsa for their patience, counsel, and support during the design and development of this book.

CREDITS

AIBO ERS-110 image courtesy of Sony USA

Angled Measuring Cup image courtesy of OXO International

Ball Vacuum Cleaner image courtesy of Dyson Technical Services

Cameleon Stroller image courtesy of Bugaboo North America

ClearRx Pill Bottle image courtesy of Target Brands, Inc.

Egg Bird Feeder image courtesy of Jim Schatz

Forehead Thermometer image courtesy of Scott Henderson of Scott Henderson Inc.

FuBar Functional Utility Bar provided courtesy of The Stanley Works

Garbino Trash Can image courtesy of Umbra

HomeHero Kitchen Fire Extinguisher image courtesy of Arnell

Hug Salt and Pepper Shakers image courtesy of Mint Inc.

iBOT Mobility System image courtesy of DEKA Research and Development Corp

LED Watch image courtesy of Fossil, Inc.

Lil' Pup Dog Bowl image courtesy of Wetnoz International

Segway i2 image courtesy of Segway Inc.

Silent Cello SVC-200 image courtesy of Yamaha Corporation of America

Sonicare Toothbrush image courtesy of David Giuliani

Stand Mixer image courtesy of Whirlpool Corporation

THUMP 2 Sunglasses image courtesy of Oakley, Inc.

Titanium Series Padlock image courtesy of Design Continuum

TOUCH! Tableware image courtesy of KAHLA Porcelain USA

ABOUT THE AUTHORS

WILLIAM LIDWELL

Will Lidwell writes, speaks, and consults on matters of design and engineering psychology. He is particularly interested in cross-disciplinary design and the means by which organizations achieve and institutionalize innovation. Lidwell is author of *Universal Principles of Design* (Rockport Publishers) and *Guidelines for Excellence in Management* (South-Western), and is a frequent contributor to popular design magazines and journals. He lives and works out of his studio in Houston, Texas.

GERRY MANACSA

Gerry Manacsa is a designer, multimedia artist and technologist living in Houston, Texas. He broke ground with mixed-media storytelling in the early days of the consumer-oriented web, foreshadowing the rise of blogs, photo sharing, and social media with a personal multimedia journal that ran for over a decade. As the web matured, he went on to explore applications for geographically dispersed workgroups, online learning, and the emerging transition to digital books. Today, he continues to trek through technology's ragged edges, seeking the useful or the fun and applying them to wide-ranging challenges in design and creativity. In his off hours, Manacsa is a photographer and explorer, looking forward to traveling the world with his wife, Karina.

Essence
- Coldspot 50

Specialization/Generalization
- Wrench 52

Assistant
- Nail 80
- Wheelchair 82
- Cooler 150
- Chair 20 Aeron

Simplicity
- Foot 92

Culture
- foot 92
- pump 122

Advocacy
- pump 122
- Q drum 154
- XO 210

Legislator
- chair 200 (?)
- Walkman 204

universal design
- Vespa 202
- peeler 74

student
- Vespa 202 (?)
- Glock 72

technotonic
- Wii 208
- stroller (mum) 40
- Coldspot 50

form follows function
- Cup 26

Gimmicks
- Cup 26

Veblen effect
- Stroller 40
- Chair 112

Clarity
- bottle 46